FRONTIERS IN PHYSICAL ORGANIC CHEMISTRY

FRONTIERS IN PHYSICAL ORGANIC CHEMISTRY

GENNADY E. ZAIKOV
EDITOR

Nova Science Publishers, Inc.
New York

For permission to use material from this book please contact us:
Telephone 631-231-7269; Fax 631-231-8175
Web Site: http://www.novapublishers.com

LIBRARY OF CONGRESS CATALOGING-IN-PUBLICATION DATA

Frontiers in physical organic chemistry / G.E. Zaikov, editor.
p. cm.
Includes index.
ISBN 1-60021-128-3
1. Chemistry, Physical organic--Research. I. Zaikov, Gennadii Efremovich.
QD476.F76 2006
547'.13--dc22 2006008658

Published by Nova Science Publishers, Inc. ✦ New York

CONTENTS

PREFACE

The further the experiment is from the theory,
The closer it is to a Nobel Prize
Frederic J. Curie
France, XX century

A man is blissful knowing nothing.
He will not be misunderstood.
Confucius
The Ancient China

Aristotle asserted that 'among the unknown in the nature surrounding us, the most unknown thing is time, because nobody knows what time is and how it can be controlled'. Since then definite positive changes have happened in this field. Particularly, the branch of science named chemical kinetics was established, which gave people the opportunity to predict the behavior of chemical reagents with time, and disclose the inner mechanism of the interaction between particles (molecules, ions, radicals and atoms) in various chemical processes.

This collection is devoted in general to chemical kinetics as in understanding the process mechanism and, consequently, gives an opportunity to forecast the behavior of molecules and their fragments, and control the process. This collection includes articles on polymer degradation and stabilization, polymerization processes, the study of specific solid and liquid reactions from positions of chemical kinetics, quantum-chemical calculations in chemical reactions, the role of transition metals in aging and stabilization of polymers, regulation of thermal conditions under fast chemical reactions, the study of the properties and practices of plastics, natural and synthetic rubbers, and the features of liquid phase structures.

Prof. Gennady E. Zaikov
N.M. Emanuel Institute of Biochemical Physics
Russian Academy of Sciences
4 Kosygin str., Moscow 119991, Russia

In: Frontiers in Physical Organic Chemistry
Editor: G.E. Zaikov, pp. 1-10
ISBN 1-60021-128-3

Chapter 1

USING THE ELEMENT SULFUR AS A STABILIZER FOR POLYMERS OF VINYLCHLORIDE

Rinat M. Akhmetkhanov[1], Natalya N. Kabalnova[2], Ruslan R. Akhmetkhanov[1], Sergey V. Kolesov[1] and Gennadi E. Zaikov[3]
[1]The Bashkir State University, 32 Frunze st., Ufa, 450074, Russia
[2]Institute of Organic Chemistry URC RAS, 71, Prospect Oktyabrya, Ufa, 450054, Russia
[3]Institute of Biophysical Chemistry RAS, 4 Kosygin st., Moscow, 119991, Russia

ABSTRACT

The high stabilizing efficiency of element sulfur was revealed by the thermooxidative destruction of polymers of vinylchloride (VC). Is was shown, that the stabilizing efficiency of sulfur in compositions of polymers of VC comparable with the efficiency of phenolic antioxidants. The element sulfur inhibits process of autooxidation of esters - plastifiers of polymers of VC, apparently, because of destroying of formed hydroperoxides.

Keywords: PVC, copolymers of vinylchloride, element sulfur, stabilizing efficiency, destruction.

INTRODUCTION

To increase thermal and thermooxidative stability of polymers VC as heat stabilisers and antioxidants some disulphides and sulpfur derivants of dialkyl tin are used [1,2]. Element sulfur, as chemical - additive to polymers, except for as curing agent for rubbers [3] till now has not found applications, though in the scientific literature there is an information about the ability of element sulfur to inhibit thermooxidative destruction of polyethylene because of destruction of hydrooxides [4]. That is why it was expedient to study the influence of element sulfur on the process of thermal and thermooxidative dehydrochlorination of hard and plasticized polyvinylchloride and some copolymers of VC.

EXPERIMENTAL PART

Thermal and thermooxidative dehydrochlorination of the hard and plasticized polymers ov VC carried out at 175^0 C in a bubbling type reactor in a current of nitrogen or air (3,3 l/h). The rate of dehydrochlorination and deacetylations were determined as in [5]. Destruction of copolymer VC with 11,1 % mol vinyl acetate (VCVA-11,1) estimated on total speed of elimination HCl and CH_3COOH. Definition of melt flow rate (MFI) of compositions was carried out on the device IIRT on a standard technique. Titration of hydroperoxides was carried out according to [6]. The polymers VC (PVC C7059M, copolymer VCVA-11,1 and copolymer VC with 28,3 % mol of methyl acrylate VCMA-28,3) were purified by washing with ethanol in a Soxhlet apparatus. Ester plastifiers dioctyl phthalate (DOP) and dioctyl sebacate (DOS) elutriated through a column filled by Al_2O_3.

In conditions of the thermooxidative destruction of polymers VC in the presence of sulfur alongside with the process of dehydrochlorination or deacetylation of polymer the formation of sulphur dioxide is possible which can influence the results of acid-base titration and by that to in to increase the rate of destruction of polymers. However, in model experiment by oxidation of individual sulfur in an atmosphere of air the qualitative test on discoloration of a solution $KMnO_4$ by oxidates of sulfur at 175^0 C at least during 80 mines, has yielded negative result. The oxidates of sulfur (175^0 C) by barbotage through distilled water did not decolourize the indicator of acid-base titration (mix of bromkrezol green and methylic red). The products of the thermooxidative decomposition of sulfur (175^0C) don't also discoloured the indicator paper kongo-rot. On the basis of obtained results it is possible to ascertain, that in conditions of carrying out of the experiment the sulphur dioxide was not formed and the obtained results on the rates of dehydrochlorination of polymers VC are correct.

RESULTS AND DISCUSSION

The introduction of the sulfur in PVC leads to the insignificant decrease of the rate of thermal dehydrochlorination of polymer (fig. 1), thus the kinetic dependences of process of elimination of HC1 have a linear kind. By an indicator method till with kongo-rot is shown, that the sulfur in conditions of destruction of PVC has no accept action in relation to allocated HCl.

By thermooxidative destruction of PVC the stabilizing efficiency of sulfur grows and particular surpasses efficiency of known antioxidant - diphenylolpropan (DPP) (fig. 1). The kind of the kinetic curves of process thermooxidative dehydrochlorination of PVC in the presence of sulfur is linear.

The maximal decrease of the rate of dehydrochlorination of PVC is observed at the contents of sulfur of the order 2 mmol/mol PVC and does not change at the further increase of the contents of sulfur in polymer. The influence of the rate of thermooxidative dehydrochlorination of PVC from the contents of sulfur after achievement of minimum rate of V_{HCl} to compare with phenolic antioxidants, for which this dependence extreme, has a linear kind. The decrease of the rate of thermooxidative dehydrochlorination of PVC in the presence of sulfur occurs practically up to meanings appropriate of the rate of thermal decomposition of PVC, that is characteristic for stabilizers - antioxidants.

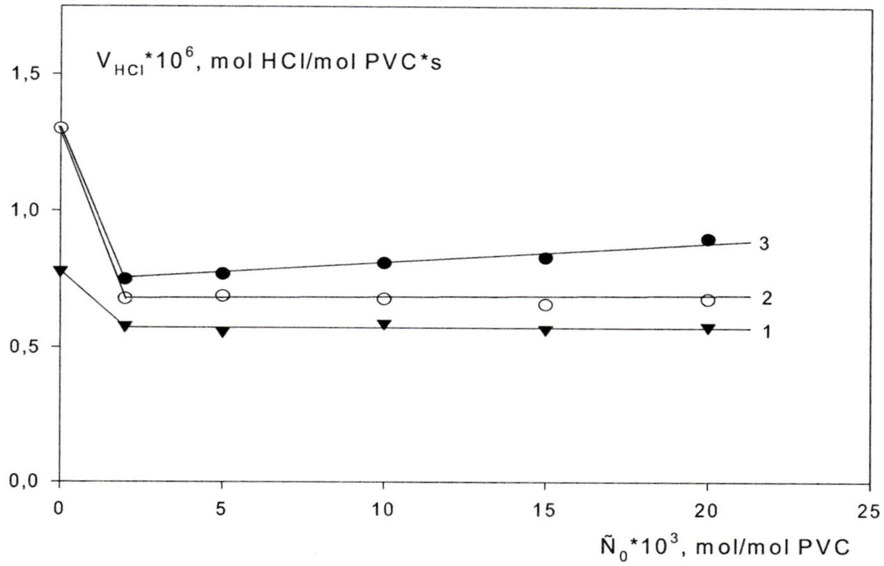

Fig. 1. Dependence of the rate of thermal (1) and thermooxidative (2,3) dehydrochlorination of PVC from the contents of elemental sulfur (1,2) and diphenylolpropan (3).

The problem of stabilization of the plasticized polymers of VC by production of soft and semifixed products appreciably connected to prevention of oxidizing destruction of plastifier in polymeric compositions by application of antioxidants.

The introduction of element sulfur in PVC, plasticized by ester-plastifiers leads to the fall off of the rate of thermooxidative dehydrochlorination of PVC (fig. 2).

The maximal decrease of the rate of dehydrochlorination of the plasticized polymer, as well as in case of destruction unplasticised PVC is observed by the contents of sulfur ≥ 2 mmol/mol PVC. The stabilizing efficiency of sulfur in PVC-plasticates practically is identical with efficiency of diphenylolpropan. The maximal stabilizing effect is reached by the introduction of sulfur in polymeric compositions in amount 0,02 weights parts/100 weights parts of PVC, that is less, than percentage by weight of known organic stabilizers - antioxidants for PVC. Thus the origin of sulfur (gas or natural) is not reflected on its stabilizing efficiency.

The decrease of rate of thermooxidative destruction of PVC by the presence of element sulfur is observed up to much lower meanings appropriate the rate of thermooxidative destruction of not plasticized PVC. Apparently, the element sulfur effectively protects plastifier from oxidation (stabilizer - antioxidant), which improves the thermo-features of PVC (known effect "an echo - stabilization" PVC [7]).

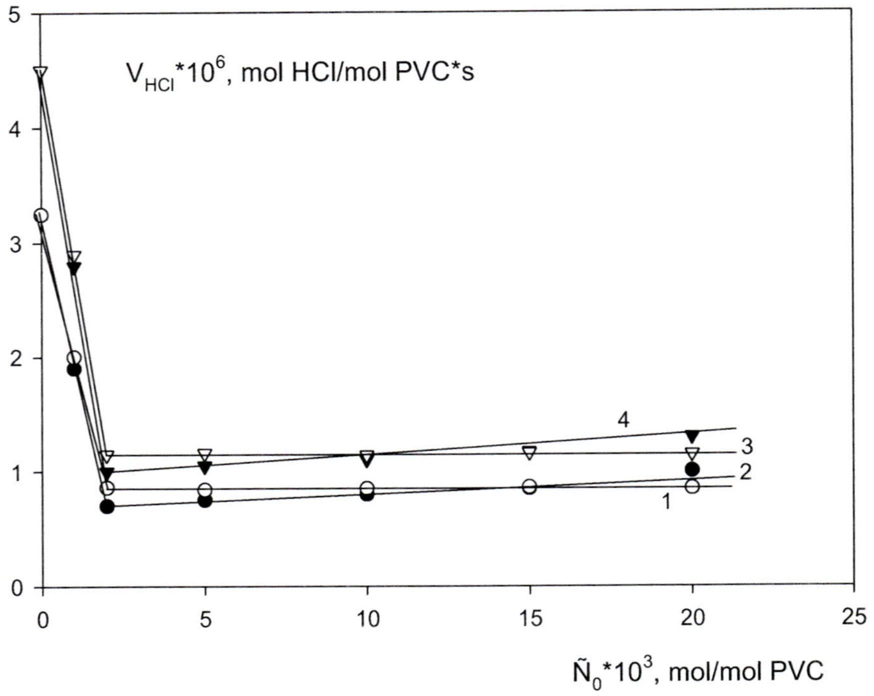

Fig. 2. Dependence of the rate of thermooxidative dehydrochlorination of PVC, plasticized (40 weights p./100 weights p. PVC) DOP (1,2) and DOS (3,4) from the contents of elemental sulfur (1,3) and diphenylolpropan (2,4).

The introduction of element sulfur in esters by their autooxidation inhibits process of accumulation hydropyroxid products. The efficiency of inhibition of process of oxidation of ester-plastifiers the element sulfur is comparable to phenolic antioxidant 4-methyl-2,6-ditertbutilphynol (fig. 3).

It is known, that the element sulfur at various temperatures passes from one modification in another [8]. Most typical transferring is observed at temperature 159-160^0 C, when soluble in carbon bisulphide S_λ, representing cyclooctasulfur S_8 turns to insoluble (polymeric) S_μ, consisting from high molecular weight circuits and rings. In an interval of temperatures 159-188^0 C the abnormal increase of melt viscosity of sulfur on four order reaching 93,2 Pa*s is observed [9] that is explained by tearing up eightparts cycles of sulfur with formation of polymeric circuits. That is why it was interesting to study the influence of sulfur on the process of dehydrochlorination of plasticized PVC at various temperatures when transferring from one modification of sulfur to another is observed.

The process thermooxidative dehydrochlorination of PVC, plasticized by dioctyl phthalate and dioctyl sebacate in the presence of element sulfur carried out at temperatures 150, 175 and 190^0 C. At 150^0 C S_λ basically is as 8-parts of rings, at 175^0 C the sulfur is intensive polimeryzed and mainly is in polymeric modification and at 190 ^{0}C it is completely in polymeric modification.

As follows from obtained datas (fig. 4) independent from the temperature, at which the destruction was carried out and independent from the modification of sulfur (elemental or polymeric) the sulfur effectively inhibits the process of thermooxidative destruction of PVC-plasticates. Stabilizing efficiency of sulfur on an index $\Delta V_{HCl}/V^0_{HCl}$ practically does not depend on temperature of destruction and reaches for PVC, plasticized by DOP and DOS accordingly 75 and 69 % (fig. 5).

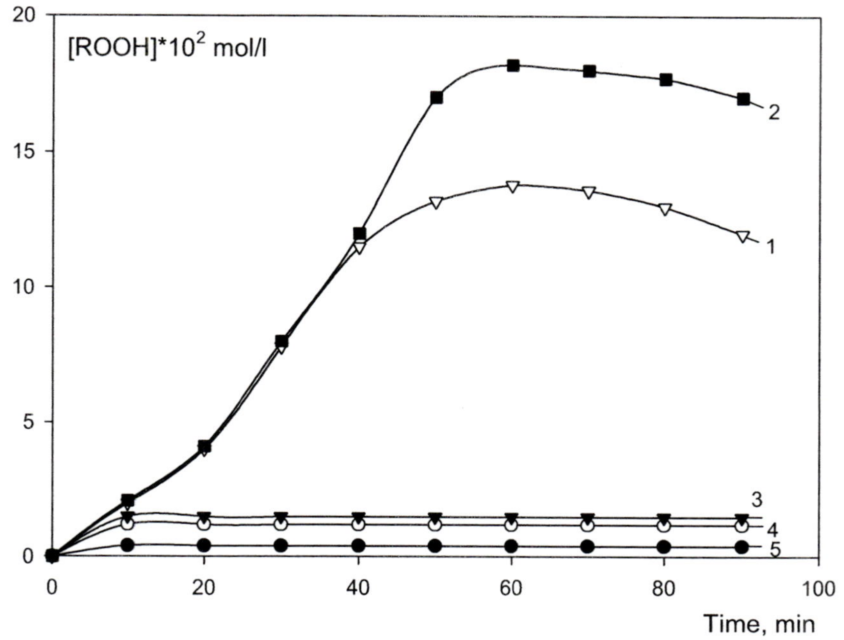

Fig. 3. The kinetic curves of accumulation of hydroperoxides by the oxidation of individual plastifiers (DOP-1, DOS-2) and plastifiers in the presence of elemental sulfur (3,1 mmol/l, DOP-3, DOS-4) and 4-methyl-2,6-ditertbutylphenol (1,4 mmol/l, DOP-5), (oxygen 3,3 l/h, 165 ^{0}C).

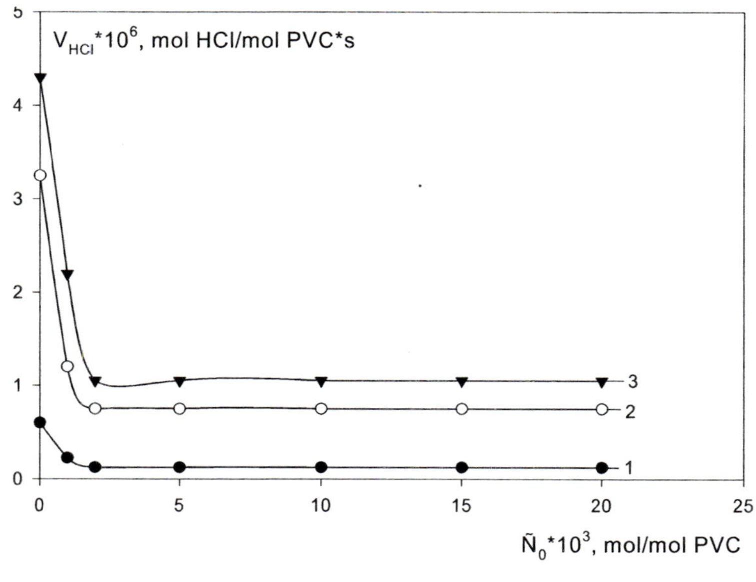

Fig. 4. Dependence of the rate of thermooxidative dehydrochlorination of PVC, plasticized 40 weights p./100 weights p. PVC by DOP from the contents of elemental sulfur at different temperatures (1-150^0 C, 2-175^0 C, 3-190^0 C).

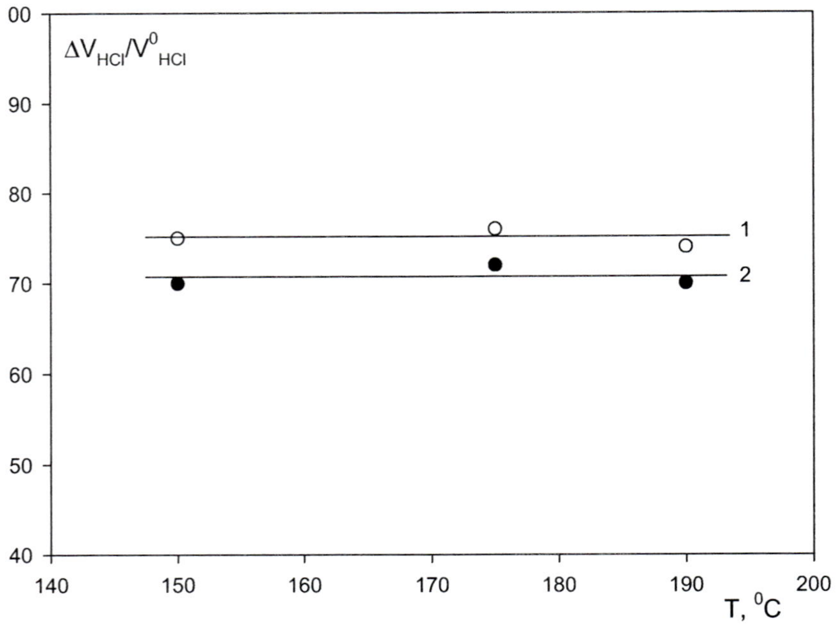

Fig. 5. Dependence of limiting stabilizing efficiency of element sulfur on an index $\Delta V_{HCl}/V^0_{HCl}$ from temperature by thermooxidative destruction (air, 3,3 l/h) PVC, plasticized by DOP (1) and DOS (2) (40 weights p./100 weights p. PVC).

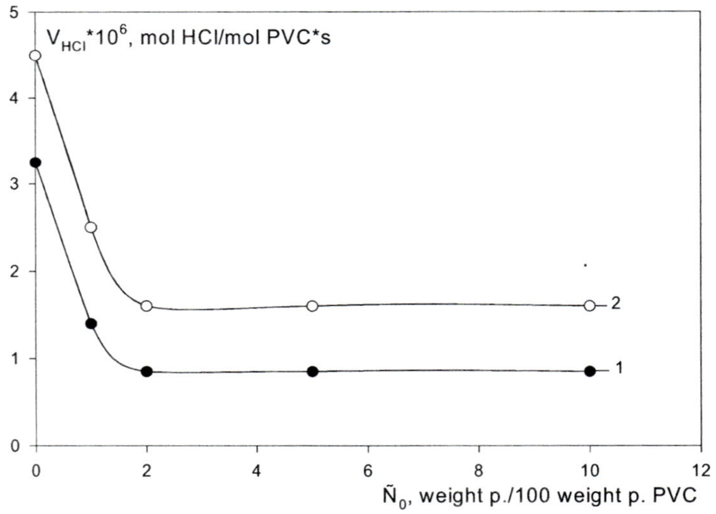

Fig. 6. Dependence of the rate of thermooxidative dehydrochloration of PVC, plasticized 40 weights p./100 weights p. PVC by DOP (1) and DOS (2) from the contents of polymeric sulfur "Crystex OT-33".

In a fig. 6 are shown the results of thermooxidative dehydrochlorination of PVC-plasticates in the presence of polymeric sulfur of industrial production "Crystex OT-33", used during vulcanization of polydienes. The polymeric sulfur as well as elemental sulfur effectively inhibits the process thermooxidative dehydrochlorination of PVC-plasticates. The kinetic laws of process of thermooxidative dehydrochlorination plasticized PVC in the presence of polymeric sulfur do not differ from the laws of inhibiting action of element sulfur.

At practical useing of small amounts of element sulfur in PVC-compositions (0,02 weights parts/100 weights parts of PVC) there is a problem of its exact dosing and uniform distribution in a polymeric composition. It is obviously possible to expedient introduction of sulfur in the plasticized PVC-compositions through its pre-dissolution in plastifier.

The limiting solubility of sulfur in dioctyl phthalate and dioctyl sebacate at 20^0 C was, accordingly, 0,41 and 0,13 % mass. For comparison the solubility of sulfur in benzol and hexane at 20^0 C makes, accordingly, 1,77 and 0,25 % of weights [9]. The limiting concentration of solutions of sulfur in esters in some times exceeds maximum effective concentration of sulfur in PVC-plasticates, that allows to dissolve the designed amounts of sulfur in plastifier, and then to mix a ready solution of sulfur in plastifier with a PVC.

Experiments have shown identical inhibiting efficiency of sulfur injected into a composition as a solution in plastifier with its efficiency at separate introduction into a composition.

The high inhibiting efficiency of element sulfur and polymeric sulfur of the mark "Crystex OT-33" in the process of thermooxidative dehydrochlorination of PVC-plasticates is confirmed in compositions containing various metalcontained heat stabilisers, on an index "time of heat stability" (tab. 1). Introduction of sulfur in amount 0,02-0,05 weights parts/100 weights parts of PVC in the stabilized PVC-plasticates increases the time of heat stability of PVC-compositions in 2,27-3,48 times depending on a nature of a metalcontained heat stabiliser, and on efficiency the sulfur does not concede to the efficiency of a phenolic antioxidant diphenylolpropan.

Table 1. The time of heat stability of PVC- compositions

Composition, weights p./100 weights p. PVC, index	1	2	3	4	5	6	7	8	9	10	11	12
PVC	100	100	100	100	100	100	100	100	100	100	100	100
DOP	40	40	40	40	40	40	40	40	40	40	40	40
Sulfur	-	0,02	-	-	0,02	-	-	0,02	-	0,02	-	0,02
DPP	-	-	0,1	-	-	0,1	-	-	-	-	-	-
Tribasic lead sulfate	2	2	2	-	-	-	-	-	-	-	-	-
$CaSt_2$	-	-	-	2	2	2	-	-	-	-	-	-
$CdSt_2$	-	-	-	-	-	-	2	2	-	-	-	-
$PbSt_2$	-	-	-	-	-	-	-	-	2	2	-	-
$BaSt_2$	-	-	-	-	-	-		-	-	-	2	2
Time, min (175^0C)	106	257	252	22	50	54	32	93	43	150	27	85

By using of element sulfur as the stabilizer – the increase of the thermooxidative of stability of compositions and the the change of melt flow index (MFI) of PVC- compositions (tab. 2) are observed. Introduction of sulfur in amounts up to 0,05 weights parts/100 weights parts of PVC leads to the reduction of MFI compositions, by the contents of sulfur more than 0,05 weights parts/100 weights parts of PVC the increase MFI compositions is observed.

Table 2. Dependence of melt flow rate of PVC-plasticates from the contents of sulfur

Composition, weights p./100 weights p. PVC, index	1	2	3	4	5	6	7	8	9	10
PVC	100	100	100	100	100	100	100	100	100	100
DOP	40	40	40	40	40	-	-	-	-	-
DOS	-	-	-	-	-	100	100	100	100	100
Tribasic lead sulfate	2	2	2	2	2	2	2	2	2	2
Sulfur	-	0,02	0,05	0,1	1,0	-	0,02	0,05	0,1	1,0
MFI, g/10 min, (190 ^{0}C)	2,34	1,78	1,46	2,03	3,70	2,80	2,60	2,45	2,75	4,10

The reduction of MFI occurs because of the presence of sulfur, that apparently, leads to the increase of molecular weight of polymer because of interaction of sulfur with internal olefinic linkages in PVC. As concentration of internal olefinic lincages in PVC is small (10^{-4} mol/mol PVC [10]) complete cross-link of polymer with loss of fluidity does not occur. At the large contents of sulfur its surplus plays a role of lubricant, increasing thus meaning MFI.

The introduction of elemental sulfur in copolymers VC from 11,1 % mol of vinyl acetate (VCVA-11,1) and from 28,3 % mol of methyl acrylate (VCMA-28,3) leads to the decrease of the rate of their thermooxidative destruction (fig. 7).

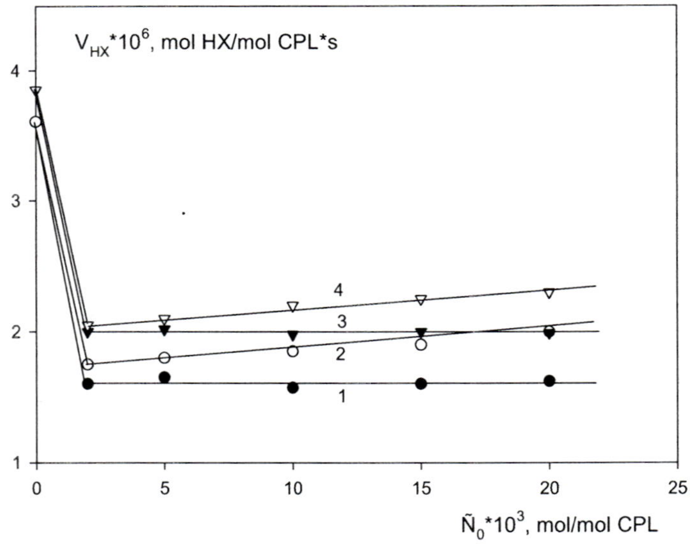

Fig. 7. Dependence of the rate of thermooxidative dehydrochlorination of VCMA-28,3 (1,2) and dehydrochlorination + deacetylation VCVA-11,1 (3,4) from the contents of element sulfur (1,3) and diphenylolpropan (2,4)

As well as in case of stabilization of thermooxidative destruction of homopolymer - PVC with the element sulfur maximal decrease of the rate of dehydrochlorination and deacetylation of copolymers VC is observed by concentration of sulfur 2 mmol/mol of copolymer. The inhibiting efficiency of sulfur in copolymers VC is also comparable with the efficiency of antioxidant diphenylolpropan (fig. 7).

The introduction of element sulfur in plasticates on the basis of copolymers VC containing metalincluded heat stabilisers, improves their thermostability on an index "time of heat stability" (tab. 3).

Table 3. Meanings of the time of heat stability of compositions on the basis of copolymers VC

Composition, weights p./100 weights p. CPL, index	1	2	3	4	5	6	7	8	9	10	11	12	13
VCVA-11,1	100	100	100	100	100	100	100	-	-	-	-	-	-
VCMA-28,3	-	-	-	-	-	-	-	100	100	100	100	100	100
DOP	40	40	40	40	40	40	40	40	40	40	40	40	40
Tribasic lead sulfate	-	-	-	-	2	2	2	-	-	-	2	2	2
$CaSt_2$	2	2	2	2	-	-	-	2	2	2	-	-	-
Sulfur	-	0,02	0,05	-	-	0,02	-	-	0,02	-	-	0,02	-
DPP	-	-	-	0,1	-	-	0,1	-	-	0,1	-	-	0,1
Time, min (175 ^0C)	10	15	16	18	19	35	31	12	18	22	16	21	24

However, it is necessary to note, that the efficiency of sulfur by inhibition of thermooxidative destruction of copolymers VCVA-11,1 and VCMA-28,3 is much lower in comparison with its efficiency of inhibition of destruction of a PVC as on an index "speed of destruction", and on an index "time of heat stability". This fact deals with the known law of reduction of effect of inhibition of stabilizers in copolymers VC in comparison with a PVC [11].

Thus, the element sulfur is the effective stabilizer - antioxidant of process thermooxidative destruction hard and plasticized polymers VC and it is interesting for practical use by production of polymeric materials.

REFERENCES

[1] Minsker K.S., Fedoseeva G.T. *Destruction and stabilization of a PVC.* Moscow, Khimiya (1979)

[2] Gorbunov B.N., Gurevich Y.A., Maslov I.P. *Chemistry and technology of stabilizers of polymeric materials.* Moscow, Khimiya (1981)

[3] Bloh G.A. *Organic vulcanization accelerators and vulcanizing systems for elastomers.* Leningrad, Khimiya (1978)

[4] Hawkins W.L., Sautter H. *Chem. and Ind. 42* 1825-1826 (1962)

[5] Minsker. K.S., Abdullin M.I., Kraikin V.A. *Plastich. Massy. 3* 31-33 (1980)

[6] Antonovsky V.L., Buzlanova M.M. *Analytical chemistry of organic pyroxid compounds*. Moscow, Khimiya (1978)

[7] Minsker K.S., Abdullin M.I. *Dokl. Acad. Nauk SSSR 263:1* 140-143 (1982)

[8] Kuznetsov A.A., Kulikova O.A. *Kauchuk I Rezina 4* 40-45 (1999)

[9] *Chemical encyclopedia*. The Large Russian Encyclopedia. Moscow V. 4. (1995)

[10] Minsker K.S., Kolesov S.V. Zaikov G.E. *Ageing and stabilization of polymers on a basis vinylchloride*. Moscow, Nauka (1982)

[11] Minsker K.S., Kolesov S.V., Pancheshnokova R.B., Akhmetkhanov R.M. *Dokl. Acad. Nauk SSSR 266:2* 370-373 (1982)

In: Frontiers in Physical Organic Chemistry
Editor: G.E. Zaikov, pp. 11-23

ISBN 1-60021-128-3

Chapter 2

OZONOLYSIS OF POLYBUTADIENES WITH DIFFERENT MICROSTRUCTURE IN SOLUTION

Metody P. Anachkov*[*]*, Slavcho K. Rakovsky, Raina V. Stefanova and Gennadii E. Zaikov**

Institute of Catalysis, Bulgarian Academy of Sciences, Acad. G. Bonchev bl.11, 1113 Sofia, Bulgaria

**Institute of Biochemical Physics, Russian Academy of Sciences, Kosigin str. 4, Moscow 119991, Russia.

ABSTRACT

The reaction of ozone with 1,4-cis-polybutadiene (E-BR) and polybutadiene having the following linking of the butadiene units (BR): 1,4-cis (47%), 1,4-trans (42%), 1,2 (11%) was investigated in CCl_4 solution. It was determined by means of IR-spectroscopy and ^{1}H-NMR spectroscopy that the basic ozonolysis products of both elastomers are ozonides and aldehydes. The aldehyde:ozonide ratio was 11:89 and 27:73 for E-BR and BR, respectively. In addition, epoxide groups were detected only in the case of BR, their yield was about 10 % of that of the aldehydes. Based on BR ozonolysis it was established that the ozonide yield from 1,4-trans units is about 50 %. By using the aldehyde yields, an evaluation was made of the efficiency of ozone degradation of the two polybutadienes, according to which the respective value of BR is considerably higher than that of E-BR. A reaction mechanism is proposed, which explains the formation of the identified functional groups and the differences in the ozone degradation of the studied elastomers.

Keywords: Ozonolysis, Polybutadiene, Solution, Ozone degradation

INTRODUCTION

It is well known that there exist some differences in the kinetics and in the mechanism of ozonolysis of the respective cis and trans alkenes [1-3]. It is established that as a rule the values of the rate constants for the trans-alkenes are about 1.5-2 times higher than those of the cis-forms [4-9]. On the other side the yield of the basic reaction product in the so called "non-participating solvents" – ozonide is usually greater in the case of cis-alkenes [2, 10, 11]. The ozone reaction with polydienes in solution is used for determination of the elastomers structure [12, 13], their modification [14] and especially for investigations on the kinetics and mechanism of ozone degradation [15, 16]. The degradation, observed in solution, is due to the proceeding of non-ozonide routes of ozonolysis, which are accompanied by formation of terminal carbonyl groups [17-19]. Our comparative study of the ozonolysis of 1,4-cis and 1,4-trans – polyisoprenes did not find considerable differences in the yields of ozonides and carbonyl compounds, and in the efficiency of ozone degradation of the two elastomers [20].

In the present paper an attempt is made to examine some peculiarities of the ozonolysis of cis- and trans- polymeric double bonds of two polybutadienes with different microstructure.

EXPERIMENTAL

Materials

Ozone

Ozone was prepared by passing oxygen through a 4-9 kV discharge.

Rubbers

Commercial samples of 1,4-cis-polybutadiene rubber (SKD) and polybutadiene (Diene 35 NFA) were used in the experiments. The molecular masses of the elastomers were determined by the GPC method on a Waters instrument. The following values of weight and number average molecular masses, respectively, were obtained: 413 000 and 154 700 for the SKD (E-BR)-samples; 298 000 and 113 300 for the Diene 35 NFA (BR) [21]. All rubbers were purified by a threefold precipitation from CCl_4 solutions with excess of methanol.

The microstructure of the polymers (in %), determined from respective ^{1}H-NMR spectra, was as follows SKD (1,4-cis->96); Diene 35 NFA (1,4-cis – 47; 1,4-trans – 42; 1,2 – 11).

PROCEDURES AND APPARATUS

Ozonation of Polymer Solutions

The experiments were performed in a bubbling reactor, equipped with a thermostatic water jacket and a viscometer. An ozone-oxygen mixture with ozone concentration in the $(0.1\text{-}1.2)\times10^{-4}$ mol/litre range was passed through the reactor at a rate of $(1.5\pm0.15)\times10^{-3}$ litre/s. The ozone concentrations in the gas phase at the reactor inlet and outlet were measured spectrophotometrically at 254 nm [3]. The volume of the polymer solutions was 15 ml for the viscometric and 10 ml for the rest of measurements. The changes in the molecular mass during ozonolysis were registered viscometrically. The algorithm for determination of the characteristic viscosity and of the degradation efficiency, defined by the number of viscometrically detected chain scission acts with respect to one molecule of reacted ozone, is described elsewhere [22]. All the viscometric measurements were performed at 25 °C.

Infrared spectra were acquired on a Carl Zeiss Specord IR-71 dispersive spectrophotometer. *1H-nuclear magnetic resonance spectra* were recorded on a Bruker WM 250 MHz instrument.

RESULTS AND DISCUSSION

Viscometric Study of Ozone Degradation

The ozone bubbling through polybutadienes solutions results in a decrease in their viscosity and, thus, in the corresponding molecular masses (Fig. 1). It is seen that the degradation of BR is more intensive. The number of chain scissions per molecule of reacted ozone (φ), derived from Fig. 1, was used to measure the efficiency of degradation of the two polybutadies towards the consumed ozone.The obtained values at $[O_3]$ = 1.2 10^5 mol/l for SKD and Diene 35 NFA were 0.007 and 0.009, respectively. It has been found that at a degree of ozone conversion of the double bonds higher than 20-25 % a gel formation is registered [22]. For that reason the ozone degradation of 1,4-cis-polybutadiene has been described as a competitive process of degradation and crosslinking routes [22]. However, it has been determined that the relative part of the crosslinking reactions in the overall balance of the consumed ozone is very low compared with the share of the degradation routes, especially under conditions of relatively low values of $[O_3]$ [17].

The φ values, determined by viscometric measurements, do not correspond to the real extent of the degradation efficiency of the elastomers [16, 19, 22, 23]. The discrepancy between the viscometric data and those obtained by functional group analysis is a result of uneven distribution of the degradation process in the bubbling reactor. The characteristics of the mass transfer of reagents during ozonolysis in bubbling reactor were discussed in detail in Ref. [16, 22]. In the case of 1,4-cis-polybutadiene, because of the very high values of its rate constant with ozone (6.10^4 $l.mol^{-1}.s^{-1}$ [2]), the instantaneous volume of the polymeric solution layer, in which the reaction takes place is about 1.1 10^{-3} cm^3 (0.007 % from the total volume) [23]; the frequency of replacement of the boundary surface of the liquid phase is about 250 times per second [22]. The intensive progress of the reaction in a very small volume determines the formation of significant amount of low molecular mass fragments which in practice cannot be viscometrically detected.

The examination of the differences in the φ values for E-BR and BR should be based mainly on two factors: (i) the conditions of mass transfer of the reagents, and (ii) the reaction mechanism. Under other identical conditions the mass transfer parameters are determined by the rate of the reaction. It is usually assumed that the 1,4-trans double bonds react with ozone about two times faster than the respective 1,4-cis-double bonds [3-9]. In this connection the rate of the ozone reaction with BR can be evaluated based on the already mentioned suggestion about respective k value and 42 % content of the 1,4-trans double bonds. Our estimation is that the influence of the relatively small difference in the rate values of E-BR and BR could not be significant. As far as the reaction mechanism is concerned, it will be shown further that in the case of BR, the relative share of the routes responsible for degradation in the overall balance of consumed ozone, is considerably greater, in comparison with the respective one of E-BR.

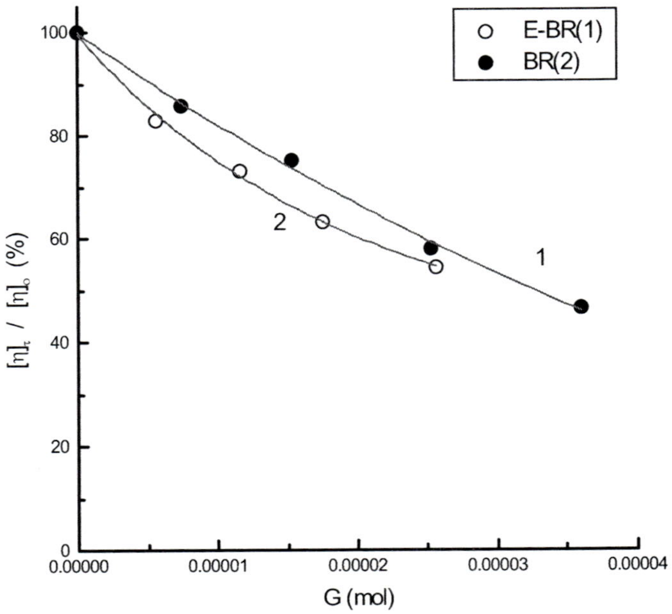

Fig.1 Dependence of $[\eta]_t / [\eta]_o$ (%) on the amount of reacted ozone (G) for carbon tetrachloride solutions (0.6 g/100 ml) of E-BR and BR. Note that $[\eta]_o$ and $[\eta]_t$ are the intrinsic viscosity values prior to and during ozonization, respectively; ozone concentration at the reactor inlet was $5x10^{-5}$ mol/litre.

Functional Group Analysis

IR-spectra of non-ozonized and ozonized polybutadiene solutions are shown in Figs. 2 and 3. In the spectra of the ozonized polybutadienes the appearance of bands at 1111 and 1735 cm $^{-1}$, that are characteristic of ozonide and aldehyde groups, corespondingly, is observed [2, 3, 17]. It was found out that the integral intensity of ozonide peak in the E-BR spectrum, is greater and of the aldehyde one is considerably smaller in comparison with the respective peaks in the BR spectrum, at one and the same ozone conversion degree of the double bonds. The mentioned differences in the aldehyde yields indicate that, according to IR-analysis, the degradation efficiency of the BR solutions is greater.

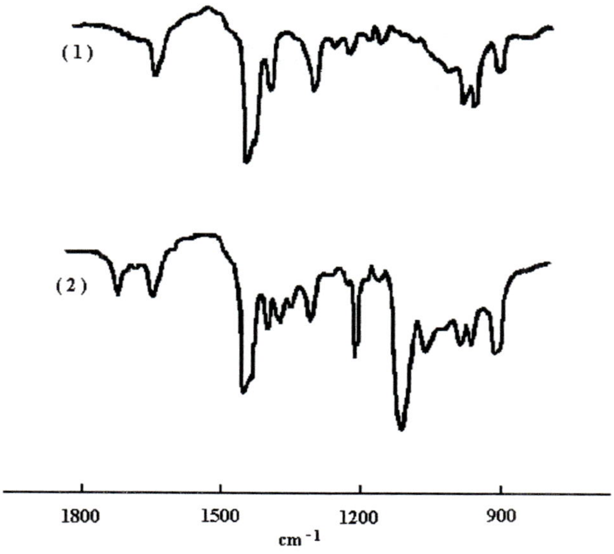

Fig. 2. IR-spectra of E-BR solutions (0.89 g / 100 ml CCl_4): (1) non-ozonized; (2) ozonized to 18 % conversion of the double bonds

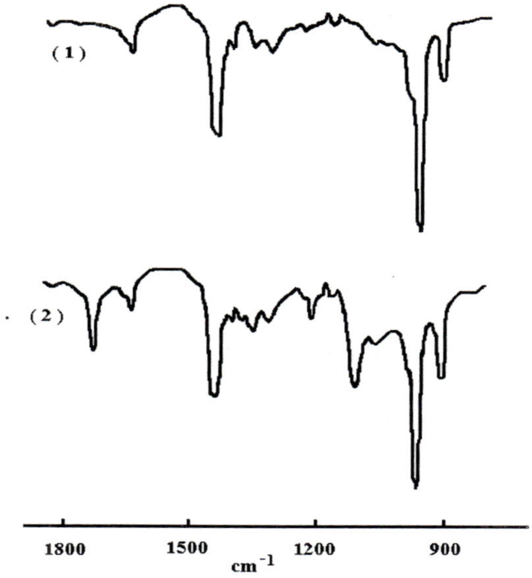

Fig. 3. IR-spectra of BR solutions (0.89 g / 100 ml CCl_4): (1) non-ozonized; (2) ozonized to 18 % conversion of the double bonds

The ^{1}H-NMR spectroscopy provides much more opportunities for identification and quantitative determination of functional groups, formed during ozonolysis of polybutadienes. Figs. 4 and 5 show spectra of ozonized E-BR and BR, respectively. The signals of the ozonolysis products are decoded in Table 1 on the basis of Fig. 6. The ozonide: aldehyde ratio, determined from NMR spectra, was 89:11 and 73:27 for E-BR and BR, correspondingly. The peak at 2.81 ppm is present only in the spectra of ozonized Diene 35 NFA. It is usually associated with the occurrence of epoxide groups [26]. The integrated intensity of that signal compared to the signal of aldehyde protons at 9.70-9.79 ppm was about 10 %. Similar signal at 2.75 ppm has been registered in the spectra of ozonized butadiene-nitrile rubbers, where the 1,4-trans double bonds are dominant [23].

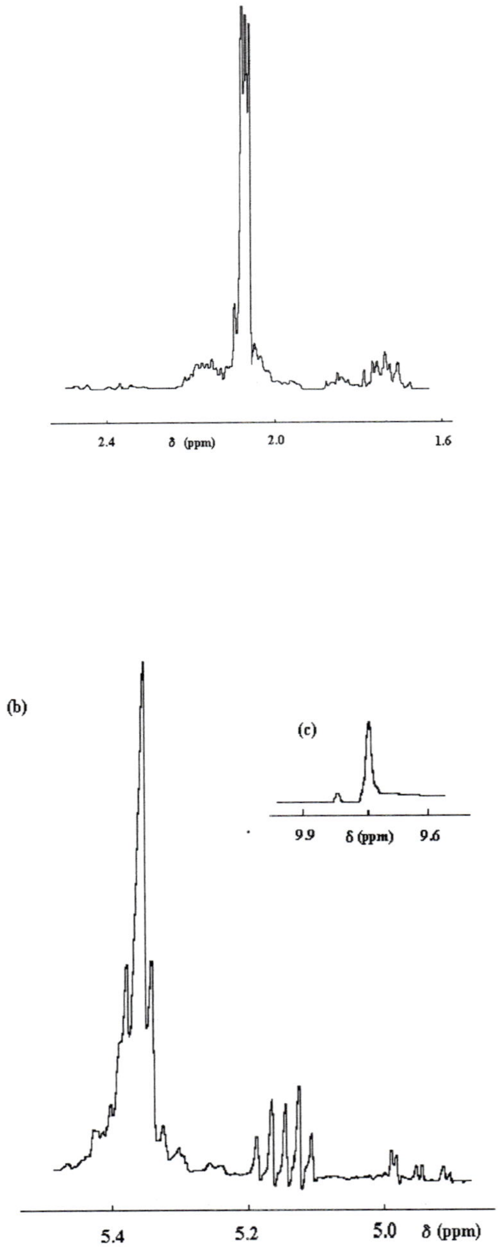

Fig. 4. ^{1}H-250 MHz NMR spectra of E-BR solutions (0.89 g / 100 ml CCl_4) ozonized to 18 % conversion of the double bonds (external standard TMS; digital resolution 0.4 Hz, 20°C).

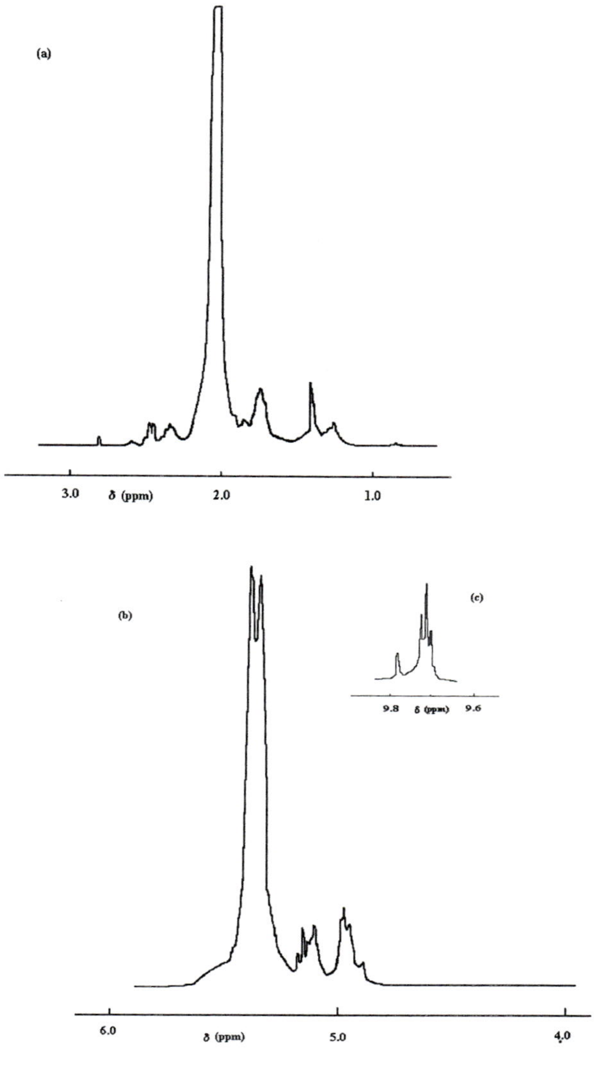

Fig. 5. ^{1}H-250 MHz NMR spectra of BR solutions (0.89 g / 100 ml CCl_4) ozonized to 18 % conversion of the double bonds (external standard TMS; digital resolution 0.4 Hz, 20°C).

There exist two isomeric forms of 1,2,4-trioxolanes [2, 10]. Their ratio is a function of the double bond stereochemistry, the steric effect of substituents and the conditions of ozonolysis and it has been studied only for low molecular weight alkenes [10, 27]. The ^{1}H-NMR spectroscopy is the most universal method for determination of the cis/trans ratio of ozonides (in the case of polymers it is practically the only method that can be applied). The measuring is based on the differences in the chemical shifts of the methine protons of t he two isomers: the respective signal of the cis form appears in lower field compared to the trans one [2, 10]. In accordance with the above mentioned, the multiplet on Fig. 3 in the area of 5.1-5.18 ppm could be interpreted as a result of partial overlapping of triplets of trans- and cis-ozonides: 5.12 (t, J≈5 Hz, 2H) and 5.16 (t, J≈5 Hz, 2H), respectively. It is interesting to note that the cis/trans ratio of the E-BR 1,2,4-trioxolanes is practically equal to that obtained from cis-3-hexene [10, 25]. The resolution of the respective BR spectrum does not allow consideration in detail of the multiplicity of the signals at 5.10 and 5.15 ppm. In this case the area of the signals is widened, most probably due to the presence of ozonide signals of the 1,2-monomer units [28].

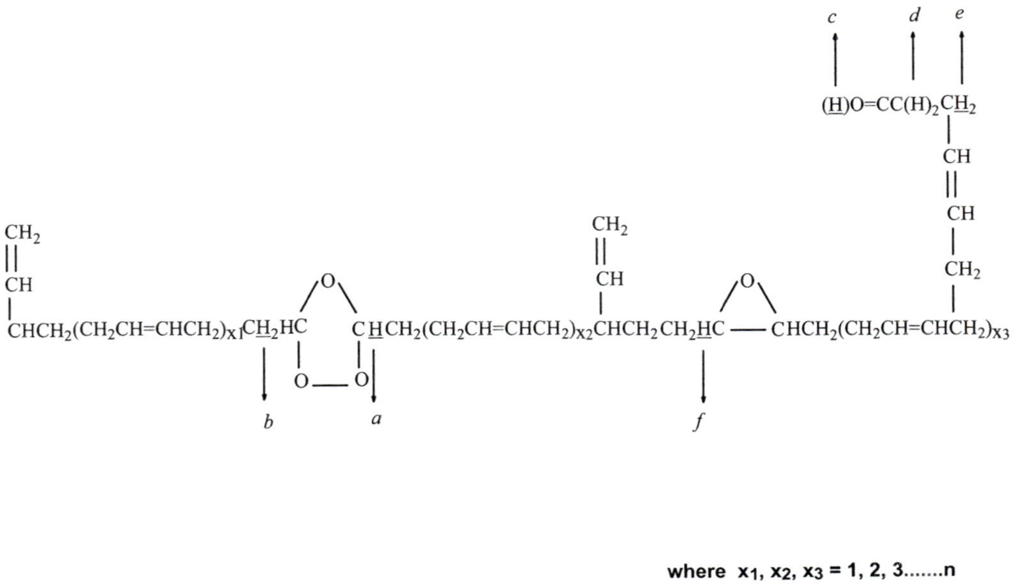

Fig. 6. Selection of protons with characteristic signals in the ^{1}H-250 MHz NMR spectra of partially ozonized polybutadiene macromolecules.

Table 1. Assignment of the signals in the ^{1}H-NMR spectra of partially ozonized E-BR and BR rubbers

Assignment of the signals (according Fig. 4)	Chemical shifts (ppm)		Literature
	E-BR	BR	
a	5.10 - 5.20 max 5.12, 5.16	5.05 - 5.18 max 5.10, 5.15	[12, 26]
b	1.67 - 1.79 max 1.72, 1.76	1.66 - 1.80 max 1.73	[12, 26]
c	9.75	9.74	[27]
d	2.42 - 2.54 max 2.47	2.42- 2.54 max 2.50	[12, 26]
e	2.27 - 2.42 max 2.35	2.27 - 2.42 max 2.35	[12, 26]
f		max 2.81	[27]

REACTION MECHANISM

The modern concepts about the mechanism of the reaction of ozone with C=C double bonds in solution are summarized in Schemes 1 and 2 [1, 2, 27-29]. The initial reaction product, the cis or trans molozonide (MO) is unstable and at temperatures higher than -150 °C for the cis and - 90 °C for the trans-isomer is decomposed to aldehyde and zwitterionic species termed as Criegee's intermediate or carbonyl oxide (hereafter referred to as CI) (Scheme 1, reaction 2) [27, 28]. The basic product of double bonds ozonolysis in non-participating solvents – the ozonides are obtained in reaction of CI with a carbonyl compound. It has been found out that the dominant part of ozonides is formed through interaction between carbonyl oxide and the corresponding carbonyl group, which both originate from the decomposition of one and the same MO, i.e. a solvent cage effect is acting (Scheme 1, reactions 3 and 3') [1, 27]. The yield of the so-called normal ozonides from the simple olefins are usually over 70 % from the total ozonide yield [27].

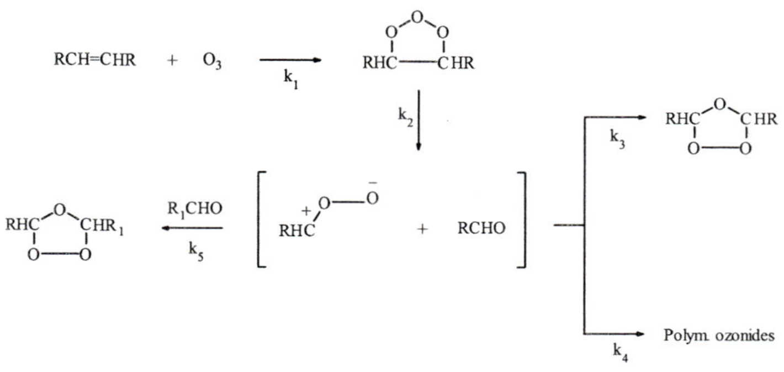

Scheme 1

Scheme 1. Formation of ozonides.

The cross-ozonide formation, observed with alkenes [1, 2, 10, 11], is possible with polydienes too (Scheme 1, reaction 5). The problem is that the application of ^{1}H-NMR spectroscopy, in the case of polybutadienes, does not allow differentiation between normal, cross and polymeric ozonides. Our experimental conditions (respective values of [C=C], [O_3] and temperature) do not favour the proceeding of reaction 5 [2]. It is obvious that the influence of the polymeric nature of the double bond substituents on the cross-ozonide formation also cannot be positive.

As it was already mentioned the ozonide yield of E-BR is considerablyy higher in comparison to that of BR. Upon investigationg of the alkene ozonolysis in solution it has been found out that the yield of normal and cross 1,2,4-trioxolanes depends on such factors as stereometric structure of the double bonds, nature of the substituents and of the solvent, temperature and alkene concentration [2, 10, 25]. There exists a general rule that the cis-alkenes usually give a higher yield of ozonides (total) and of normal 1,2,4-trioxolanes than the trans isomers [2, 10]. For example the maximal ozonide yields of the cis- and trans 3-hexene are respectively 88 and 49 %, and the minimum values are 81 and 22 % [2, 10]. In the case of cis and trans 2-butene the maximum values are 72 % and 36 % and the minimum 48

% and 36 % correspondingly [2, 10]. The high selectivity of the reaction towards 1,2,4-trioxolanes is also typical of the ozonolysis of the terminal double bonds which could be considered as analog of the 1,2- monomer units [28]. The analysis of the reaction selectivity towards the ozonides shows similarity between the 1,4-cis-polybutadiene units and the cis-3-hexene, while in the case of cis-2-butene the results are lower. An attempt was made to evaluate the yield from 1,4-trans units, based on their contents in BR, the data determined by ^{1}H-NMR for E-BR and assuming the ozonide yield of 1,2-units equal of that of 1,4-cis units. We obtained that the selectivity towards ozonides for the 1,4-trans units of polybutadiene is about 50 %.

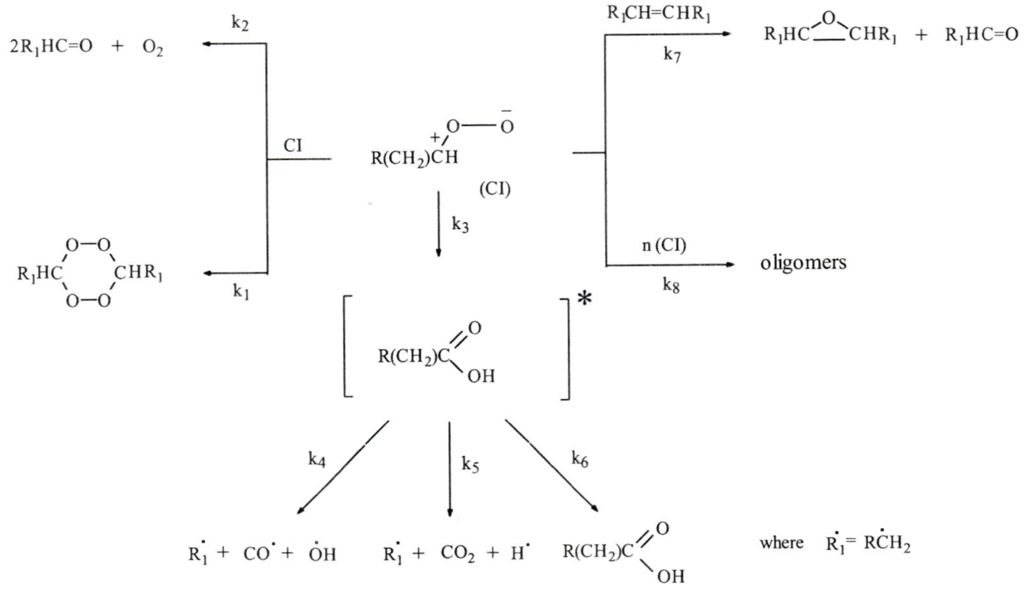

Scheme 2

Scheme 2. Alternative routes of conversion of CI intermediates.

As it was reported previously at 20-25 % ozone conversion of polybutadienes, especially of those with high contents of 1,4-cis units a gel formation is observed [17, 22]. It has been found out that the crosslinks of polymers containing double C=C bonds in the side chain have peroxide nature [30]. The kinetics of the crosslinking process during ozonolysis of polymers with double bonds has been studied in [23, 31]. Our expectations are that in the case of the polybutadienes, the predominant part of crosslinks are result of formation of polymeric ozonides (Sheme 1, reaction 4), but their share in the overall balance of reacted ozone is very small, of the order of a few percents.

The second main product of the polybutadiene ozonolysis are the aldehydes. According to the mechanism of the ozone reaction with double bonds in solution, the aldehyde groups are being formed when the conversion of CI intermediates is proceeding through routes, which are an alternative to the carbonyl oxide-aldehyde interaction [2, 29]. The non-ozonide routes are illustrated in Scheme 2. In the case when the CI conversion results in formation of carbonyl group, two moles of aldehydes are obtained from one mol of ozone (Scheme 2, reactions 2 and 7). Another route of CI deactivation is its isomerization via hot acid to radicals (Scheme 2, reactions 4, 5). In this case 1 mol of ozone produces 1 mol of aldehyde [3]. Acidic groups were not detected among the reaction products (Scheme 2, reaction 6). The stability and the life time of low molecular CI intermediates with different substituents and their capability to undergo various monomolecular and bimulecular interactions are widely discussed in the literature [28, 29]. The peculiarities of conversion of two intermediates in polyisoprene ozonolysis (one of them is the same as that obtained from polybutadiene) are considered in Ref. 20. It has been shown that, under our experimental conditions, the dominant 'non-ozonide' route of conversion of the respective carbonyl oxide intermediate is its isomerization via hot acid.

The epoxide groups may be formed as a result of direct interaction between ozone and a double bond, or by a reaction of a double bond with dioxirane, which could be considered as the cyclic form of the carbonyl oxide intermediate [28, 29]. However there is no evidence that the above mentioned reactions or the cyclization of the respective intermediate are taking place with alkenes, which are structural analogues of the polybutadienes, or with the very polybutadienes. The most probable suggestion is that the epoxide groups are products of interaction of the carbonyl oxide with a double bond (Scheme 2, reaction 7). It should be noted that, in contrast to the ozonolysis of cis- and trans-hexene [25], 1,4-cis- and 1,4-trans-polyisoprene [20], in the case of polybutadienes the formation of epoxide groups is observed only when the elastomer macromolecules contain 1,4-trans monomer units. This phenomenon could be explained with the relatively higher stability of the dominant stereochemical form (syn carbonyl oxide) of the respective intermediate that originates from ozonolysis of 1,4-trans units [29].

The basic route of the reaction - the formation of normal ozonides does not lead directly to a decrease in the molecular mass of the elastomer macromolecules, because the respective 1,2,4-trioxolanes are relatively stable at ambient temperature [31, 32]. The most favourable conditions for ozone degradation emerge when the cage interaction (Scheme 1, reaction 3) does not proceed. Therefore the higher the ozonide yield the lower the intensity of ozone degradation of the polybutadienes and vice versa. As it was already determined the ozonide yields for the 1,4-cis- and 1,2- monomer units are of the order of 83-90 %, whereas that for the 1,4-trans units is about 50 %. The amount of aldehyde groups is usually used for evaluation of the intensity and efficiency (number of chain scissions per molecule of reacted ozone) of ozone degradation of elastomers. In this case it should be taken into account that the dominant route of degradation leads to the formation of 1 mol of aldehyde from 1 mol of ozone.

CONCLUSIONS

The reaction of ozone with 1,4-cis-polybutadiene and polybutadiene containing 1,4-cis-; 1,4-trans- and 1,2-monomeric units was investigated in CCl_4 solution. The aldehyde:ozonide ratio was 11:89 and 27:73 for E-BR and BR, respectively. In addition, epoxide groups were detected, only in the case of BR and their yield was about 10 % of that of the aldehydes. Based on the BR ozonolysis, it was established that the ozonide yield from 1,4-trans units is about 50 %.

The non-ozonide routes of deactivation of the Crigee's intermediates were discussed. A supposition is substantiated that monomolecular decomposition to radicals via isomerization to hot acid is the dominant route for CI conversion. Taking into account the aldehyde yields, an evaluation was made of the efficiency of ozone degradation of the two polybutadienes, according to which the respective value of BR is considerably higher than that of E-BR.

REFERENCES

[1] R. Crigee, *Angew. Chem. Int. Ed. Eng.* 1975, 14, 745.

[2] P. S. Bailey, *Ozonation in organic chemistry*, Vols 1, 2, Academic Press, New York 1978, 1982.

[3] S. D. Razumovskii; G. E. Zaikov, *Ozone and its reactions with organic compounds*, Elsevier, Amsterdam 1984.

[4] S. Rakovsky; G. Zaikov; *Kinetics and mechanism of ozone reactions with organic and polymeric compounds in liquid phase*, Nova Science Publishers Inc., New York, 1998.

[5] D. G. Williamson, R. J. Cvetanovic, *J. Am. Chem. Soc.* 1968, 90, 4248.

[6] W. Pritzkow, G. Schappe, *J. Prakt. Chem.* 1969, 311, 689.

[7] D. Cremer, *J. Am. Chem. Soc.* 1981, 103, 3619.

[8] Ch. R. Greene, R. Atkinson, *Int. J. Chem. Kinet.* 1992, 24, 803.

[9] S. D. Razumovskii, *Khimicheskaja Physika*, 2000, 19 (7), 58.

[10] R.W. Murray, R.D.Youssefyeh, P. R. Story. *J. Am. Chem. Soc.* 1967, 89, 2429.

[11] J-I. Choe, M. Sprinivasan, R. L. Kuczkowski, *J. Am. Chem. Soc.* 1983, 105, 4703.

[12] M. J. Hackathorn, M. J. Brock, *Rubber Chem. Technol.* 1972, 45, 1295.

[13] I. Spasskova, *Vestn. Leningr. Univ., Khimia*, 1973, No 10, 140.

[14] G. G. Egorova, V. S. Shagov, "Ozonolysis in the Chemistry of Unsaturated Polymers", in: *Synthesis and Chemical Transformation of Polymers*, Izd. LGU, Leningrad, 1986. (In Russian).

[15] Yu. S. Zuev, T. G. Degtereva, *Elastomer stability under operating conditions*, Khimia, Moscow, 1986 (In Russian).

[16] D. Bruck, *Kautch.+Gummi Kunststoffe*, 1989, 42, 760.

[17] M. P. Anachkov, S. K. Rakovsky, D. M. Shopov, S. D. Razumovskii, A. A. Kefely, G. E. Zaikov, *Poly. Deg. and Stab.* 1985, 10, 25.

[18] G. Ivan, M. Giurginka, *Poly. Deg. and Stab.* 1998, 62, 441.

[19] F. Cataldo, Polym. Degrad. Stab. 2001, *73*, 511.

[20] M. P. Anachkov, S. K. Rakovsky, R. V. Stefanova, *Poly. Deg. and Stab.* 2000, 67, 355.

[21] M. P. Anachkov, Unpublished results.

[22] S. D. Razumovskii, M. P. Anachkov, A. A. Kefely, G. E. Zaikov, *Visokomolek. Soed. Ser. B*, 1982, 24, 94.

[23] M. P. Anachkov, S. K. Rakovsky, R. V. Stefanova, D. M Shopov, *Poly. Deg. and Stab.* 1987, 19, 293.

[24] M. P. Anachkov, S. K. Rakovsky, R. V. Stefanova, A. K. Stoyanov, *Poly. Deg. and Stab.* 1993, 41, 185.

[25] R. W. Murray, W. Kong, S. N. Rajadhyaksha, *J. Org. Chem.* 58, 1993, 315.

[26] *The Sadtler Handbook of Proton NMR Spectra*, Sadtler, Philadelphia, PA, 1978.

[27] R. L. Kuczkowski, *Chem. Soc. Reviews*, 1992, 79.
[28] K. J. McCullough, M. Nojima, "Peroxides from Ozonization", in: *Organic Peroxides*, Ando W., Ed., John Wiley and Sons Ltd, 1992.
[29] W. H. Bunnelle, *Chem. Rev.* 1991, 91, 335.
[30] M. P. Anachkov, S. K. Rakovsky, D. M. Shopov, S. D. Razumovskii, A. A. Kefely, G. E. Zaikov, *Poly. Deg. and Stab*. 1986, 14, 189.
[31] M. P. Anachkov, S. K. Rakovsky, A. K. Stoyanov, J. Appl. Polym. Sci., 1996, 61, 585.
[32] M. P. Anachkov, S. K. Rakovsky, Bulg. Chem. Comm. 2002, 34 (3/4), 486.

In: Frontiers in Physical Organic Chemistry
Editor: G.E. Zaikov, pp. 25-40
ISBN 1-60021-128-3

Chapter 3

HYDRIDE ADDITION OF METHYLHYDRIDSILOXANES TO 1,3-CYCLOHEXADIENE

O. Mukbaniani*[1,*], *G. Zaikov*[2], *U. Scherf*[3], *S. Meladze*[1], *T. Tatrishvili*[1], *Z. Pachulia*[3], *Kh. Koberidze*[1] and *M. Labartkava
[1] I. Javakhishvili Tbilisi State University,
I. Chavchavadze Ave., 1, Tbilisi 380028, Georgia
[2]Institute of Biochemical Physics Russian Academy of Sciences
4 Kosygin Street, Moscow 119991 (Russia)
[3]Makromolekulare Chemie Bergische Universitat Wuppertal
Gauss-Str. 20, D-42097 Wuppertal

ABSTRACT

The reaction of hydride polyaddition of methylhydridesiloxanes to 1,3-cyclohexadiene, at the ratio of initial compounds 1:30, in the presence of platinum chlorohydric acid has been investigated. The hydride polyaddition reaction order, activation energies and rate constants were found. The polyorganosiloxanes with 1,3-cyclohexadiene fragments in the side chain, completely soluble in organic solvents were synthesized. By quantum-chemical half-empirical method AM1 the electronic characteristics of 1,3-cyclohexadiene: the charges on the carbon atoms and bonds orders were calculated. For fully characterization of hydride addition of methylhydridesiloxane to 1,3-cyclohexadiene, by quantum-chemical half empiric AM1 method, for all initial, intermediate and final products, in modelling reaction of hydrosilylation of methyldimethoxysilane to 1,3-cyclohexa-diene, the heats of formations (ΔH_f), change of energy (ΔH) of the system depending on the change of distance ($R_{C\text{-}Si}$) between ≡C-Si≡ bonds, also the charges values (q) on the atoms, dipole moments (μ) and bonds orders (Pij) are calculated. It was concluded, that the course of hydride addition of modelling reaction of methyldimethoxysilane to 1,3-cyclohexadiene energetically is more favourrable by 1,4-addition. ^{1}H and ^{13}C NMR, IR, thermogravimetric, gel-permeation

* To whom correspondence should be addressed, E-mail: mukbaniani@ictsu.tsu.edu.ge

chromatographic, differential scanning calorimetric and X-ray methods were characterized the synthesized oligomers.

Keywords: organosiloxanes, hydride polyaddition, oligomers thermal-oxidative stability.

INTRODUCTION

The properties of siliconorganic polymers besides of other factors depend as on the structure of a macromolecular chain and on the nature of the surrounding organic groups at silicon atom [1].

The effective methods of synthesis of new type siliconorganic copolymers with complex valuable properties attach a great importance. Particularly it's very interesting the synthesis of polyorgano-siloxanes with unsaturated surrounding groups at silicon atom, for receive siliconorganic rubbers [2] and modifiers for carbo chain polymers [3] is very important.

For this point of view the chemical modification reactions are very interesting. In literature the mo-dification reactions of the linear polymethylvinylsiloxanes with norbornene groups are known [4]. In this case the synthesized polymers are characterized with higher reaction ability for vulcanisation and thermal-oxidative stability. By the catalytic dehydrocondensation of polymethylhydridesiloxa-nes with hydroxyorganocyclosiloxanes in the presence of catalysts the polyorganosiloxanes with cyclic fragments in the side chain were synthesised [5]. By hydrosilylation reactions of methyl(eth-yl)hydridesiloxane with vinylorganocyclosiloxanes in the presence of platinum hydrochloric acid polyorganosiloxanes with carboorganocyclosiloxane fragments in the side chain were synthesized [1,6]. In both cases the insertion of reactionable cyclic fragments in the side chain leads to the rise of thermal-oxidative stability and glass transition temperature of copolymers.

By polymerisation reaction of methyltricyclodecenylcyclotri(tetra)siloxanes in the presence of cata-lysts methylsiloxane copolymers with unsaturated tricyclodecenyl groups in the chain was obtained [7].

By hydrosilylation reaction of tricyclodecadiene with α,ω-bis(trimethylsiloxy)-methylhydridesilo-xanes in the presence of platinum hydrochloric acid or platinum on the carbon as a catalyst various linked methylsiloxane oligomers with tricyclodecenyl fragments in the chain were synthesized [8]:

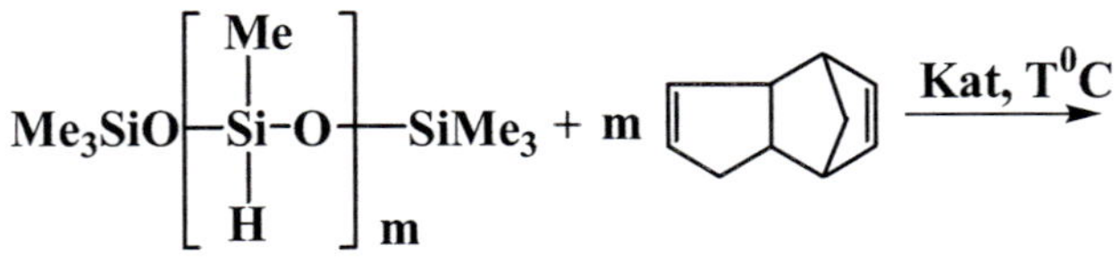

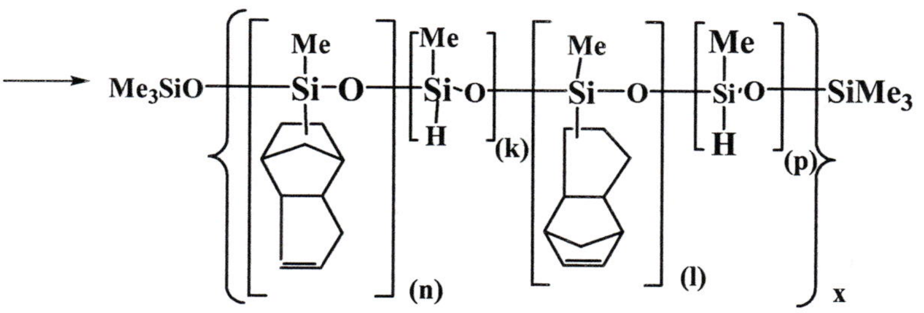

Where: [(n) + (k) + (l) + (p)]x=m; m≈30.

It was established that with the rise of temperature three-dimensional polymers are received. During hydrosilylation at 90^0C about 20% insoluble polymers were obtained, this may be explained by the secondary interchain reactions on the bases of catalyst in reaction system. Soluble oligomers were obtained in case of 80% conversion of active ≡Si-H bonds.

EXPERIMENTAL PART

The purity of initial dicyclopentadiene was controlled by its boiling temperature and refractive in-dex. The organic solvents were cleaned by drying and distillation. The purity of the starting compo-unds was controlled by gas-liquid chromatography "LKhM-8-MD" (Russian model), phase SKTF-100 (10%), the NAW chromosorb, carrier gas He, the 2M column.

The IR spectra of all the samples were taken on an UR-20 instrument, while the ^1H and ^{13}C nuclear magnetic resonance (NMR) spectra were recorded on a Bruker AC-250 spectrometer at 250 MHz in the solution of deuterium solvents as a solvent and internal standard.

The starting materials for the synthesis of the linear siloxanes with cyclohexenyl fragments in the si-de chain were α,ω-bis(trimethylsiloxy)methylhydridesiloxanes with degree of polymerization n≈30 and 1,3-cyclohexadiene.

Gel-permeation chromatography (GPC) measurements were carried out by using toluene or THF as an eluent and IR detector. Standards polystyrenes or polydimethylsiloxanes were used for calibra-tion. Gel permeation chromatography was carried out with the use of a Waters Model 6000A chro-matograph with an R 401 differential refractometer detector. The column set comprised 103 and 104 Å Ultrastyragel columns. Sample concentrations were approximately 3% by weight in toluene, and typical injection volume for the siloxane was 5μL. Standardization of the GPC was accomplished by use the of styrene or PDMS standards with the known molecular weight.

Diffractograms were taken on a "Dron-2" instrument (Russian model), A-Cuα values were measu-red without a filter, the angular velocity of the motor ω=2 dg/min.

The thermogravimetric investigations were carried out on a "Paulic-Paulic-Erdey" derivatograph model MOM-102 (Hungary). The test conditions were: temperature rise rate v=5 deg/min, in the op-en area.

Results and Discussion

In literature there is no information about cyclohexenyl side group containing linear methylsilo-xanes. The modification reactions of α,ω-bis(trimethylsiloxy)methylhydride-siloxanes (n≈30) with 1,3-cyclohexadiene for the purpose of synthesis of the linear siliconorganic elastomers with cyclo-hexenyl side groups were studied. The hydrosilylation reactions were studied at 1:30 molecular ratio of initial compounds in the presence of platinum chlorohydric acid as a catalyst. Preliminary heating of initial compounds in the temperature range of 60÷90^{0}C in the presence of catalyst showed that in these conditions no changes of initial compounds take place. Besides, there are no changes in the NMR and IR spectra of 1,3-cyclohexadiene and methylhydridesiloxanes. By gas-liquid chroma-tography it was established that the polymerisation of 1,3-cyclohexadiene in this condition does not take place.

As it's seen from the structure of 1,3-cyclohexadiene the hydrosilylation may proceed to one double bond with preservation of the second one. But another side hydride addition may be realized in "1-2" direction or as a conjugated diene system in "1-4" direction (Scheme 1).

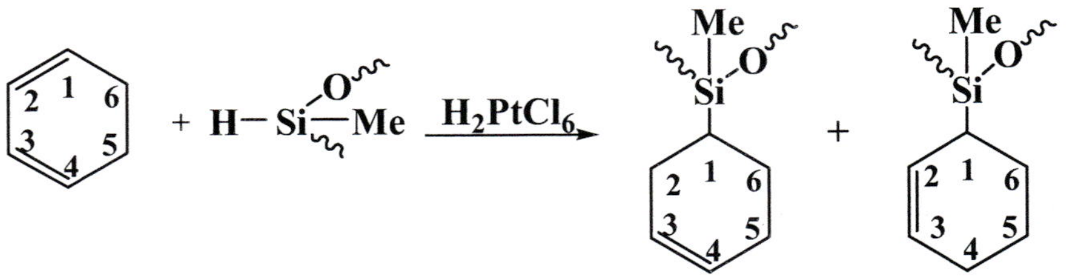

Scheme 1

By quantum-chemical half-empirical method AM1 [9] the electronic characteristics of 1,3-cyclo-hexadiene: the charges on the carbon atoms and bonds orders were calculated.

The obtained results are presented on the molecular diagram 1:

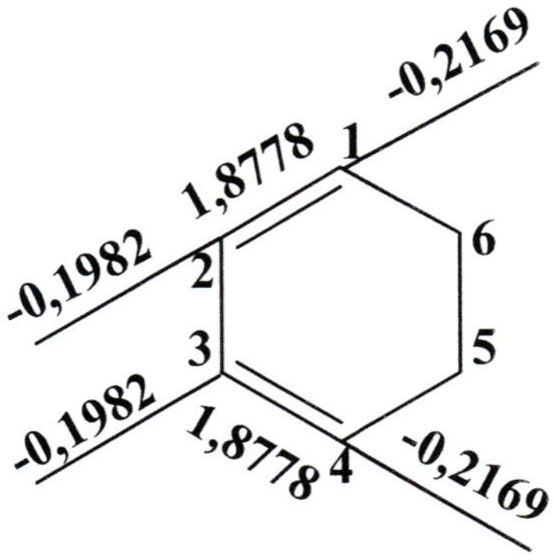

Diagram 1

As it's seen from the molecular diagram, the higher value of the charge is observed on the carbon atoms in 1 and 4 positions. This indicates that these carbon atoms are more reactionable centres for electrophilic addition. Because of that the hydride addition may be realized with more probability in "1-4" position, than 2 or 3 position ($q_{1,4}$≈–0,2169; $q_{2,3}$ ≈–0,1982). The ratio of atoms charges values $q_{1,4}/q_{2,3}$≈1,09, which is peculiarly in agreement with NMR spectra data. Besides bond orders in "1-2" and "3-4" position are characterized with the same values $P_{1,2}=P_{3,4}$≈1,8778, which indicate that during hydrosilylation decisive rule execute the charges values on the atoms.

It was established that during the hydrosilylation reaction in melt condition, at the beginning stages of hydrosilylation (20% of the conversion of active ≡Si-H groups) three-dimensional systems were obtained, which may be explained not only by intramolecular reactions, but also by intermolecular reactions of macro-chain (Scheme 2).

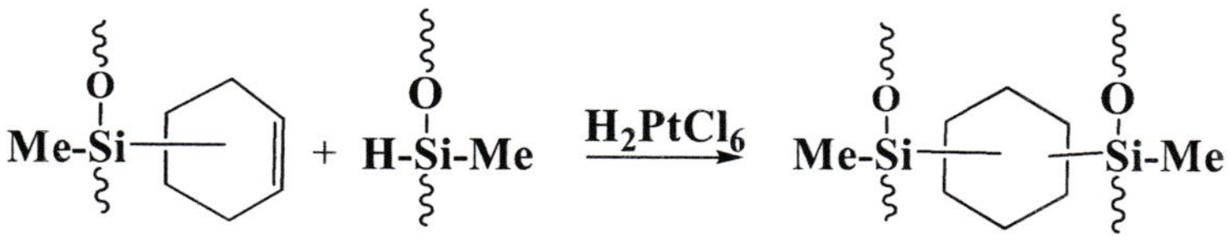

Scheme 2

For the purpose of synthesis of soluble organosiloxanes the hydrosilylation reaction were carried out in absolute toluene solution (C≈0,1155 mole/l). The course of the reaction was observed by decrease of the amount of active ≡Si-H groups. As it's seen from Figure 1, with the rise of temperature the depth of hydrosilylation (conversion of ≡Si-H groups) reaction increases.

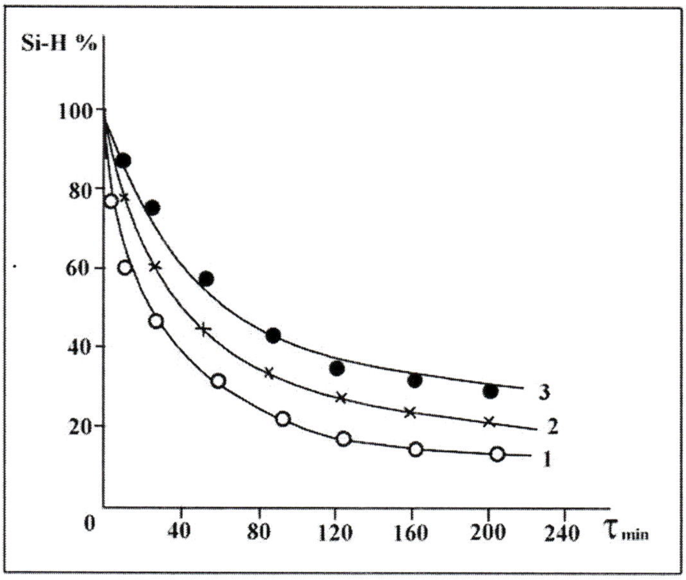

Figure 1. Dependence of changes of active ≡Si-H% groups on the time during hydride addition of α,ω-bis(trimethylsiloxy)methylhydridesiloxanes to 1,3-cyclo-hexadiene. Where curve 1 is at 80^{0}C, curve 2 is at 70^{0}C and curve 3 is at 60^{0}C.

After 3h hydrosilylation at 60^{0}C temperature the conversion of active ≡Si-H bonds is about 69% (Figure 1, curve 3). With the rise of the temperature the conversion of active ≡Si-

H bonds increases and at 80^{0}C temperature it consists about 85% (curve 1). It must be denoted that at higher stages of hydrosilylation also three-dimensional systems were obtained. For the purpose of synthesis of solub-le systems the hydrosilylation reaction put an end at ~80% depth of ≡Si-H conversion.

Consequently hydrosilylation reaction of methylhydridesiloxanes to 1,3-cyclohexadiene proceeds according to the following scheme 3.

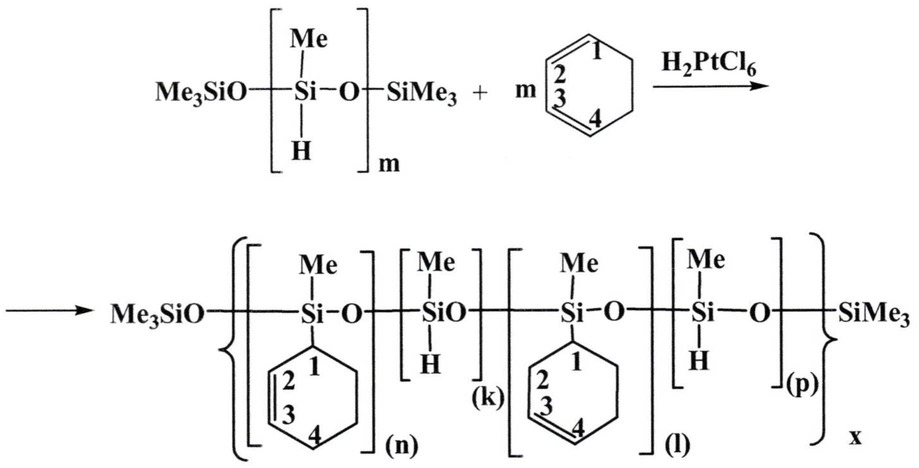

Where: [(n)+(k)+(l)+(p)]x≈30, I^1 (60^{0}C); I^2 (70^{0}C); I (80^{0}C).

Scheme 3

The synthesized siliconorganic oligomers were determined by elementary analysis, by finding of molecular masses, by IR and NMR spectra data. Elementary composition, yields and some physical-chemical properties of oligomers are presented in Table 1.

Table 1. Elementary composition, yields and some physical-chemical properties of oligomers

Oligo -mer №	Yield, %	Reaction tempera-ture, ^{0}C	η_{sp}*	T_g, ^{0}C	d_1, Å	5%-mass losses	$\bar{M}_\omega$x10^{-3}	Elementary** composition, %		
								C	H	Si
I^1	94	60	0,08	-14	8,40	230	-	59,42 58,95	8,66 8,43	20,54 21,02
I^2	94	70	0,08	-	-	-	-	59,42 59,21	8,66 8,51	20,54 20,94
I	95	80	0,09	-10	8,43	230	48	59,42 59,30	8,66 8,35	20,54 20,77

*At 25^{0}C temperature, in 1% solution of toluene. **Overline there are calculated values and under-line - experimental values.

In the IR spectra of oligomers one can see characteristic bands for asymmetric valence oscillation of the linear ≡Si-O-Si≡ bonds at 1020 cm^{-1}. In the spectra one can observe characteristic bands for ≡Si-Me, -$SiMe_3$ and ≡Si-H bonds at 1275, 840 and 2170 cm^{-1} accordingly.

In the ^1H NMR spectra (Figure 2) of synthesized oligomer I one can see the singlet signals for methyl protons with chemical shift with centre $\delta \approx 0,1$ ppm, in the spectra one can also see widened singlet signals for methyl protons in fragment -CH=CH- with chemical shift $\delta \approx 6,1$ and $\delta \approx 5,5$ ppm. From spectra it is evident that the hydrosilylation proceeds to as "1-4" direction also to "1-2" direction, exceed "1-4" direction and the ratio is ~1,6:1.

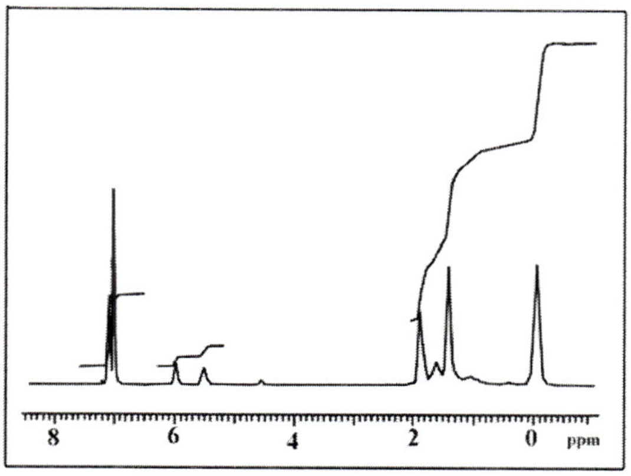

Figure 2. ^1H NMR spectra of α,ω-bis(trimethylsiloxy)methylhexenylsiloxane I. (Solvent and standard $CDCl_3$).

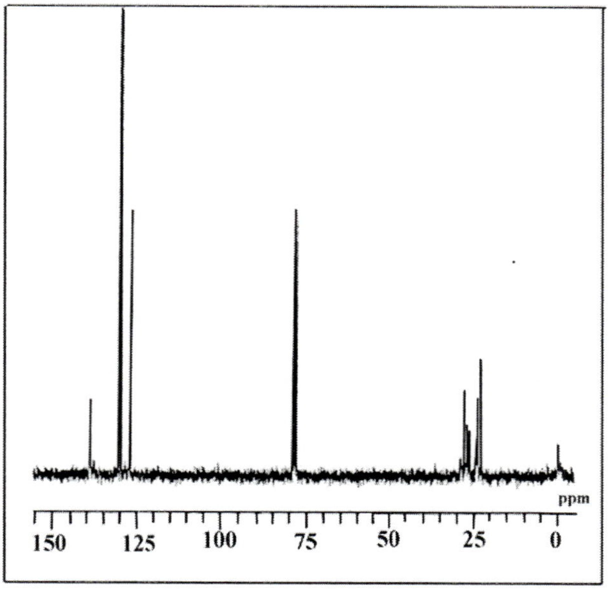

Figure 3. ^{13}C NMR spectra of α,ω-bis(trimethylsiloxy)methylhexenylsiloxane I.(Solvent and standard $CDCl_3$).

In ^{13}C NMR spectra (Figure 3) of oligomer I one can observe for "1-4" and "1-2" hydride addition signals group with chemical shift: for "1-4" addition $\delta \approx 129,6$; $\delta \approx 130,4$ ppm and for "1-2" addition $\delta \approx 126,7$ and $\delta \approx 139,9$ ppm. Accordingly chemical shifts are easily given from saturated section characterising for "1-4" addition at 23,7(C-1); 29,2(C-5); 27,8(C-4); 130,4(C-2); 23,7(C-1) and for "1-2" addition 24,1(C-6); 28,8(C-5); 129,6(C-4), 126,7(C-3); 22,8(C-2); 27,0(C-1).

For fully characterization of hydride addition of methylhydridesiloxane to 1,3-cyclohexadiene, by quantum-chemical semi empiric AM1 method [9], for all initial, intermediate and final products, in modeling reaction of hydrosilylation of methyldimethoxysilane to 1,3-cyclohexadiene, the heats of formations (ΔH_f), change of energy (ΔH) of the system depending on the change of distance ($R_{C\text{-}Si}$) between ≡C-Si≡ bonds, also the charges values (q) on the atoms, dipole moments (μ) and bonds or-ders (Pij) are calculated (see Tables 2 and 3).

For the beginning the modeling reaction of hydride addition of methyldimethoxysilane [$Me(MeO)_2SiH$] to 1,3-cyclohexadiene to 1,2-direction was considered. The model reaction proce-eds with formation of compound II according to the following scheme 4.

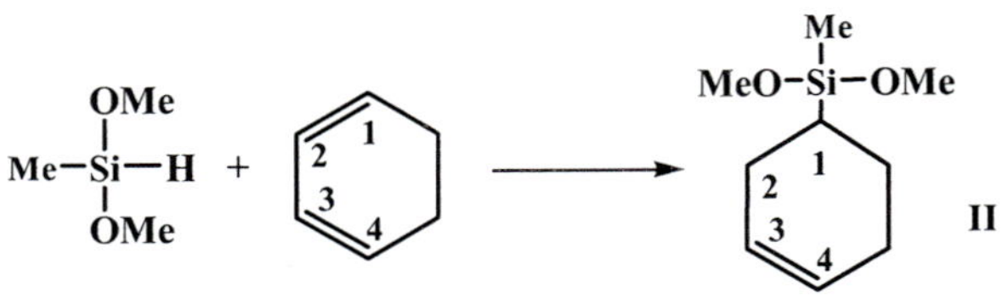

Scheme 4

The distances between the double bonded carbon atom (C_1) and silicon atoms in 1,3-cyclohexadiene was taken about 1,0 Å more than the bond, expected in a product. Figure 4 showed the starting position and numeration of atoms of compounds in modeling hydrosilylation reaction of methyldi-methoxysilane to 1,3-cyclohexadiene:

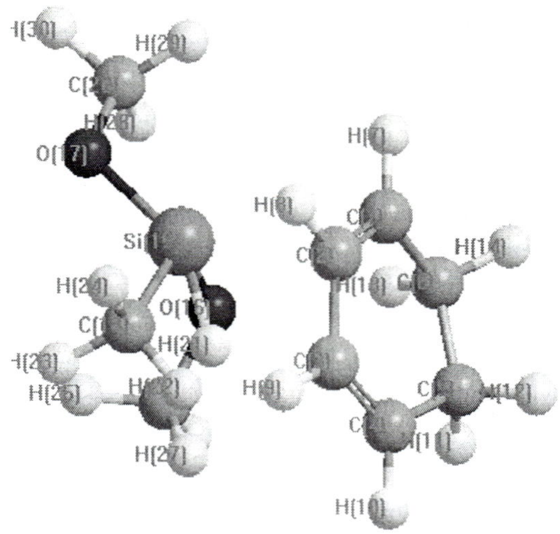

Figure 4. Starting position and numeration of atoms of compounds in modeling 1,2-hydrosilylation reaction of methyldimethoxysilane to 1,3-cyclohexadiene

The change step of distances ($R_{C_1\text{-}Si}$) between atom of silicon and double bonded carbon atom (C_1) was 0,05 Å. Dependence of change of energy (ΔH) of the system on the distances ($R_{C\text{-}Si}$) between silicon and carbon atoms is presented on the Figure 5.

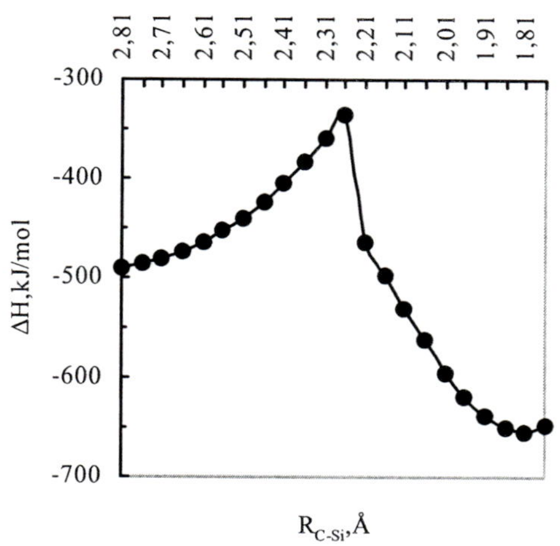

Figure 5. Dependence of changes of energy (ΔH) on distances ($R_{C_1\text{-}Si}$) between silicon and carbon atoms during of modeling hydrosilylation reaction of methyldimethoxysila-ne to 1,3-cyclohexadiene (1,2-addition).

As it is seen from Figures 5, at approximation of silicon to atom of carbon (C_1) up to $R_{C\text{-}Si}$=2,26 Å distances the energy of system rises. At the same time the bond order between atoms of carbon C_1 and C_2 in a molecule of 1,3-cyclohexadiene decreases from 1,855 up to 1,648. The bond order of between hydrogen and silicon atoms also decreases ($R_{Si\text{-}H}$=0,907-0,739) and in modeling compound methyldimethoxysilane the process of formation of new bonds ($R_{C_1\text{-}Si}$=0,013-0,106 and $P_{C_2\text{-}H}$= 0,001-0,073) is observed. In case of distance up to 2,21 Å the energy of system sharply decreases, the double C=C bond passes in ordinary C-C bond ($P_{C_1\text{-}C_2}$=0,996), and as to the bond order betwe-en silicon and carbon atom C_1 that it reaches to 0,794. Thus the hydrogen completely eliminated from silicon ($P_{Si\text{-}H}$=0,011) and attach to carbon atom C_2 ($P_{C_2\text{-}H}$=0,912).

With consecutive reduction of distances the energy of system monotonously decreases and with strengthening formed new bonds. The heat of formation of the product II of hydride 1,2-addition (ΔH=-654,04 kJ/Mole) is in a good correlation with calculated heat of formation (ΔH=-653,94 kJ/Mole, see Table 2) of compound II obtained by hydride addition of methyldimethoxysilane to 1,3-cyclohexadiene at 1,2-direction. Activation energy of 1,2-addition of modeling reaction is equal to E_{act}=155,16 kJ/Mole. On the Figure 6 one can see final structure of modeling 1,2-addition of methyldimethoxysilane to 1,3-cyclohexadiene.

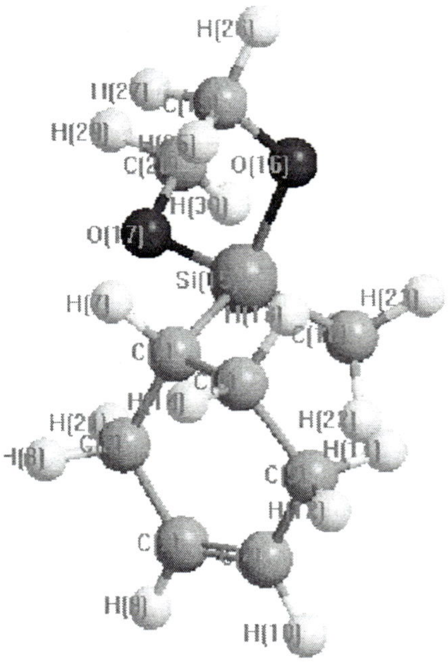

Figure 6. Final structure of the product obtained by modeling 1,2-addition of methyldimethoxysilane to 1,3-cyclohexadiene.

The 1,4-addition of modeling compound methyldimethoxysilane to 1,3-cyclohexadiene is consider-ed also. The modeling reaction of 1,4-addition proceeds with formation of compound III according to the scheme 5.

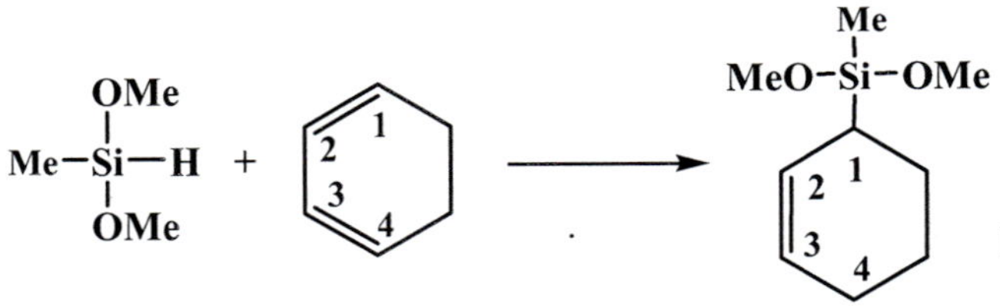

III

Scheme 5

Figure 7 showed the starting position of compounds in modeling 1,4-hydrosilylation reaction of methyldimethoxysilane to 1,3-cyclohexadiene:

On the Figure 8 the dependence of change of energy (ΔH) of the system on the distances ($R_{C\text{-}Si}$) between silicon and carbon atom is presented. Similarly the above-mentioned the distance between the double bonded carbon atom (C_1) and silicon was taken about 1,0 Å more than the bond, expec-ted in a product. The change pace of distances ($R_{C_1\text{-}Si}$) between atom of silicon and double bonded carbon atom (C_1) was about 0,05 Å.

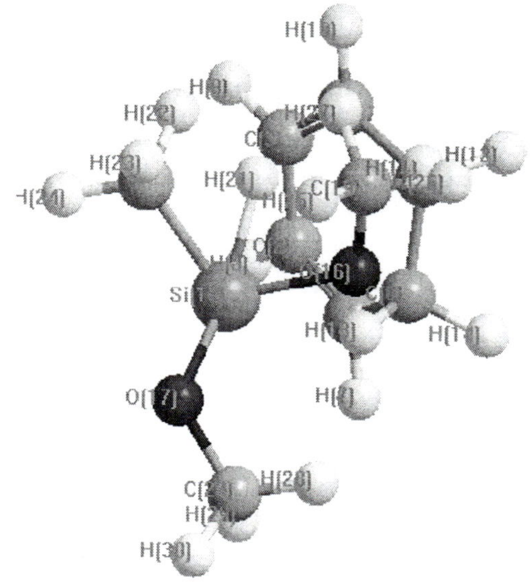

Figure 7. Starting position of compounds in modeling 1,4-hydrosilylation reaction of methyldimethoxysilane to 1,3-cyclohexadiene

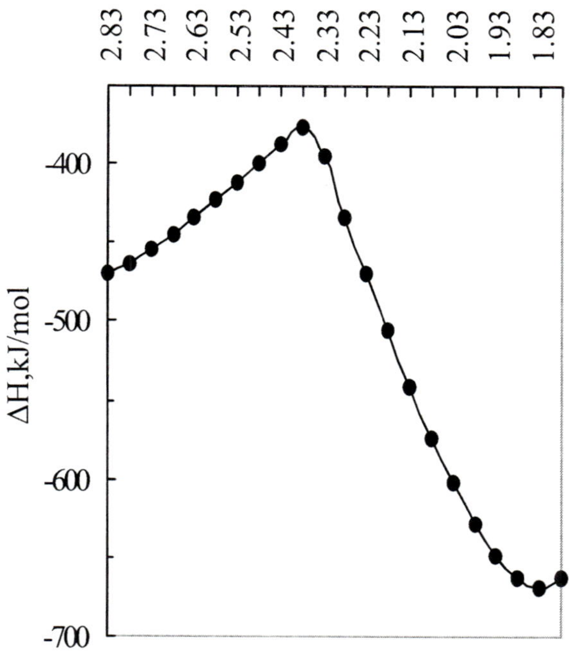

Figure 8. Dependence change of energy (ΔH) on distance ($R_{C\text{-}Si}$) between silicon and carbon atoms during modeling hydride addition of methyldimethoxysilane to 1,3-cyclohexadien by 1,4-addition.

As it is seen from Figures 8, during approximation of silicon to atom of carbon (C_1) up to $R_{C\text{-}Si}$=2,38 Å distances the energy of system also rises. Thus the bond order of between atoms of carbon C_1-C_2 and C_3-C_4 in a molecule of 1,3-cyclohexadiene decreases from 1,829 up to 1,630 and from 1,840 up to 1,627 accordingly. Thus the bond order of between atoms of carbon C_1-C_2 and C_3-C_4 in a mole-cule of 1,3-cyclohexadiene decreases from 1,829 up to 1,630 and from 1,840 up to 1,627 according-ly. In parallel the bond order between atoms of carbon C_2 and C_3 ($P_{C_2-C_3}$=1,063-1,205) is increased, and the process of formation of new bonds (P_{C_1-Si}=0,027-0,185 and P_{C_4-H}=0,014-0,170) is obser-ved. From 2,33 Å distances the reduction of energy of system is observed sharply, the double bonds C_1=C_2 and C_3=C_4 pass in ordinary C-C bonds ($P_{C_1-C_2}$=1,065 и $P_{C_3-C_4}$=1,046) and the bond order between silicon and carbon atom C_1 reaches 0,779.

Thus hydrogen completely eliminated from silicon and (P_{Si-H}=0,011) and attach to carbon atom C_4 (P_{C_4-H}=0,878). The transition of usual ordinary C_2-C_3 bond in double bond also is observed, that is accompanied by sharp change of value of bond orders between atoms C_2 and C_3 ($P_{C_2=C_3}$=1,817). On the Figure 9 one can see the structure of final product obtained by modeling 1,4-addition of methyldimethoxysilane to 1,3-cyclohexadiene.

Table 2. Change of energy (ΔH) of the system depending on the change of distance (R_{C-Si}) between ≡C-Si≡ bonds, the charges values (q) on the atoms, dipole moments (μ) and bonds orders (Pij), during modeling hydride 1,2-addition in model reaction of methyldimethoxysilane to 1,3-dicyclohexadiene

$R_{1,15}$, Å	ΔH, kJ/mol	q_1	q_2	q_{15}	q_{21}	μ	P_{12}	$P_{1,15}$	$P_{15,21}$	$P_{2,21}$
2.81	-491.11	-.1862	-.1263	1.3613	-.2266	1.214	1.855	.013	.907	.001
2.76	-485.86	-.1776	-.1401	1.3738	-.2337	1.550	1.846	.020	.897	.003
2.71	-480.11	-.1860	-.1351	1.3746	-.2327	1.457	1.843	.023	.895	.004
2.66	-472.91	-.1966	-.1282	1.3755	-.2329	1.376	1.840	.027	.893	.005
2.61	-463.94	-.2099	-.1192	1.3770	-.2335	1.296	1.835	.031	.890	.006
2.56	-452.89	-.2261	-.1079	1.3793	-.2346	1.222	1.829	.035	.886	.008
2.51	-439.50	-.2457	-.0939	1.3829	-.2365	1.161	1.821	.041	.881	.011
2.46	-423.54	-.2702	-.0762	1.3888	-.2398	1.116	1.810	.047	.873	.015
2.41	-404.91	-.3011	-.0536	1.3984	-.2453	1.108	1.794	.056	.860	.020
2.36	-383.71	-.3438	-.0221	1.4143	-.2545	1.143	1.766	.068	.839	.030
2.31	-360.45	-.3979	.0187	1.4440	-.2714	1.302	1.723	.084	.802	.045
2.26	-335.94	-.4660	.0708	1.4943	-.2962	1.474	1.648	.106	.739	.073
2.21	-464.87	-.3891	-.0820	1.3615	-.0006	1.413	.996	.794	.011	.912
2.16	-498.63	-.3997	-.0869	1.3734	.0125	1.471	.993	.799	.009	.921
2.11	-530.82	-.4119	-.0907	1.3865	.0242	1.516	.991	.802	.008	.928
2.06	-560.93	-.4258	-.0938	1.4015	.0347	1.567	.990	.805	.006	.934
2.01	-594.18	-.4030	-.1208	1.3951	.0858	.472	.987	.847	.006	.939
1.96	-618.28	-.4236	-.1180	1.4164	.0879	.480	.985	.845	.005	.943
1.91	-637.40	-.4456	-.1152	1.4387	.0897	.504	.984	.842	.004	.947
1.86	-649.96	-.4688	-.1124	1.4621	.0917	.557	.983	.838	.003	.951
1.81	-653.95	-.4936	-.1087	1.4860	.0930	.647	.982	.835	.003	.954
1.76	-646.89	-.5198	-.1039	1.5104	.0936	.771	.980	.830	.002	.957

Table 3. Change of energy (ΔH) of the system depending on the change of distance (R_{C-Si}) between ≡C-Si≡ bonds, the charges values (q) on the atoms, dipole moments (μ) and bonds orders (Pij), during modeling hydride 1,4-addition in model reaction of methyldimethoxysilane to 1,3-dicyclohexadiene

$R_{1,15}$, Å	ΔH, kJ/mol	q_1	q_4	q_{15}	q_{21}	μ	P_{12}	P_{23}	P_{34}	$P_{1,15}$	$P_{4,21}$
2.83	-469.79	-.1726	-.1349	1.3299	-.1804	1.760	1.829	1.063	1.840	.027	.014
2.78	-463.01	-.1758	-.1321	1.3233	-.1783	1.749	1.823	1.066	1.835	.032	.017
2.73	-454.27	-.1806	-.1274	1.3196	-.1756	1.648	1.815	1.070	1.827	.038	.022
2.68	-444.82	-.1862	-.1218	1.3157	-.1728	1.543	1.804	1.076	1.818	.046	.028
2.63	-434.63	-.1928	-.1151	1.3118	-.1699	1.433	1.791	1.084	1.805	.055	.037
2.58	-423.72	-.2002	-.1070	1.3079	-.1669	1.316	1.776	1.094	1.788	.068	.048
2.53	-412.15	-.2092	-.0974	1.3045	-.1636	1.206	1.755	1.107	1.767	.084	.064
2.48	-400.10	-.2191	-.0863	1.3017	-.1598	1.147	1.728	1.127	1.737	.105	.086
2.43	-387.98	-.2306	-.0730	1.3001	-.1552	1.129	1.689	1.156	1.694	.136	.118
2.38	-376.64	-.2443	-.0565	1.3007	-.1489	1.179	1.630	1.205	1.627	.185	.170
2.33	-394.47	-.2928	-.1072	1.2904	.0107	.755	1.065	1.817	1.046	.779	.878
2.28	-433.89	-.3027	-.1185	1.2977	.0327	.701	1.041	1.851	1.024	.809	.914
2.23	-470.50	-.3155	-.1215	1.3152	.0420	.672	1.034	1.861	1.019	.814	.921
2.18	-505.96	-.3295	-.1244	1.3339	.0507	.651	1.027	1.871	1.014	.819	.927
2.13	-541.80	-.3433	-.1265	1.3533	.0572	.692	1.015	1.887	1.010	.829	.933
2.08	-573.63	-.3598	-.1284	1.3733	.0639	.664	1.011	1.893	1.007	.830	.938
2.03	-602.61	-.3777	-.1299	1.3939	.0700	.633	1.008	1.898	1.004	.831	.942
1.98	-627.88	-.3968	-.1309	1.4154	.0752	.619	1.005	1.903	1.002	.830	.945
1.93	-648.31	-.4173	-.1312	1.4377	.0794	.626	1.002	1.908	1.000	.828	.949
1.88	-662.43	-.4391	-.1308	1.4605	.0827	.658	1.000	1.912	.998	.826	.952
1.83	-668.33	-.4621	-.1297	1.4839	.0851	.722	.997	1.915	.997	.823	.955
1.76	-662.17	-.4966	-.1277	1.5177	.0866	.872	.994	1.919	.996	.818	.957

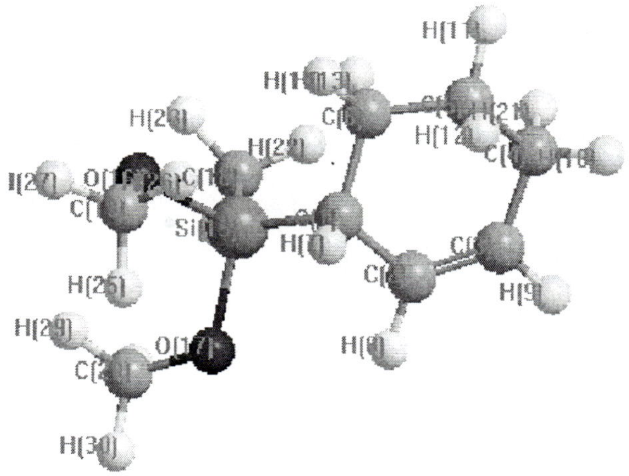

Figure 9. Final structure of the product obtained by modeling 1,4-addition of methyldimethoxysilane to 1,3-cyclohexadiene.

With consecutive reduction of distances between silicon and carbon atoms the energy of system monotonously decreases and with strengthening formed new bonds. The heat of formation of a pro-duct of modeling hydride 1,4-addition (ΔH=-656,87 kJ/Mole) is in a good correlation with the calculated heat of formation (ΔH=-668,33 kJ/Mole, see Table 3) of compound III obtained by modeling hydride 1,4-addition of methyldimethoxysilane to 1,3-cyclohexadiene. Activation energy of 1,4-addition of modeling reaction is equal to E_{act} = 93,15 kJ/Mole.

Thus at comparison of the Figures 5 and 8, and also energy of activation (the character of curve of change of energy and the value of bond orders) one can be concluded, that the course of modeling reaction of hydride addition of methyldimethoxysilane to 1,3-cyclohexadiene energetically is more favourable by 1,4-addition, that is in agreement with NMR spectral data.

On Figure 10 the dependence of reverse concentration on the time is presented, from which it is evident that at the beginning stages the hydrosilylation reaction is the reaction of the second order. From Figure 10 the reaction rate constants at various temperatures have been calculated: $K_{60^0C} \approx 1{,}4402x10^{-2}$, $K_{70^0C} \approx 2{,}330x10^{-2}$ and $K_{80^0C} \approx 3{,}2x10^{-2}$.

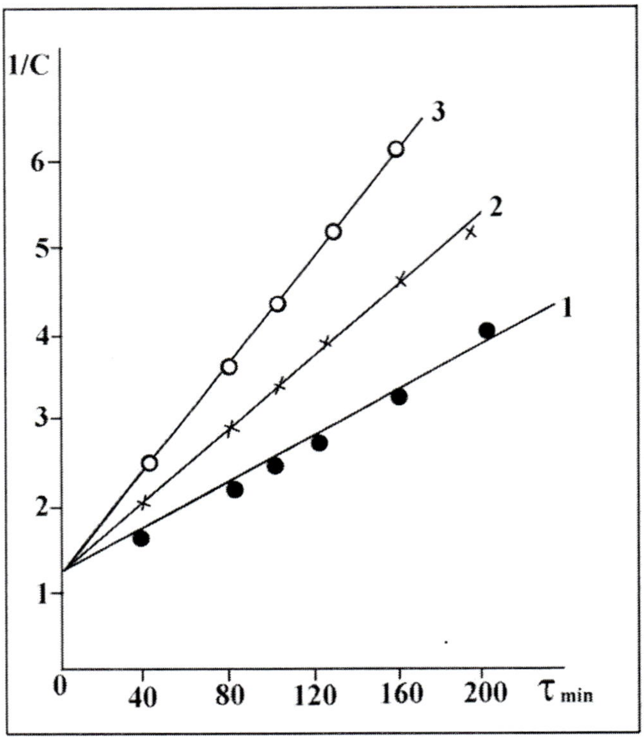

Figure 10. Dependence of reverse concentration on the time, where curve 1 is at 60^0C, curve 2 is at 70^0C and curve 3 is at 80^0C.

Figure 11 shows the dependence of logarithm of the reaction rate constants on the reverse tem-perature. From these data, the activation energy of the hydrosilylation reaction was calculated to be equal $E_{activ} \approx 35{,}0$ kJ/mole. This value of activation energy is close to the activation energy of hyd-rosilylation of dicyclopentadiene with α,ω-bis(trimethylsiloxy)methylhydridesiloxanes ($E_{activ} \approx 32{,}3$ kJ/mole).

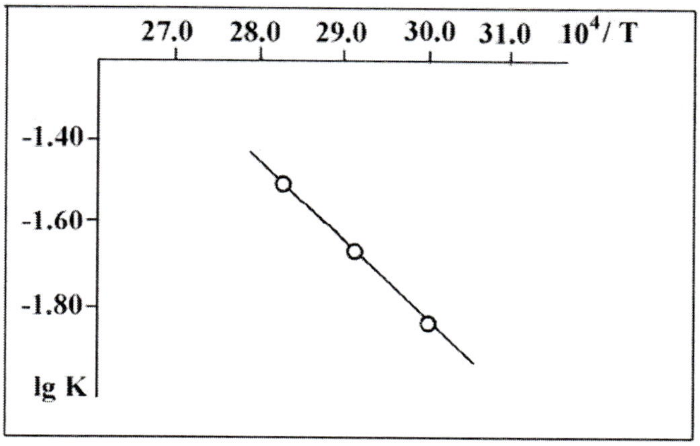

Figure 11. Dependence of reverse temperature on the logarithm of the rate constants of hydride addition of α,ω-bis(trimethylsiloxy)methylhydridesilo-xane to 1,3-cyclohexadiene.

The synthesized oligomers were studied by GPC method. Figure 12 shows molecular weight dist-ribution of oligomer I, which showed that it has a polymodal character of distribution. The average weight molecular mass of oligomer is equal to $48x10^3$. In case of full hydrosilylation the value of average weight molecular mass does not exceed $4,3x10^3$. Increased molecular mass of obtained oligomers once again proved that the reaction proceeds by macromolecular interchain hydride addi-tion with receiving of the branching oligomers. So in synthesized oligomers the elementary links have a different structure. With the rise of reaction temperature the intra- and interchain hydride addition increases.

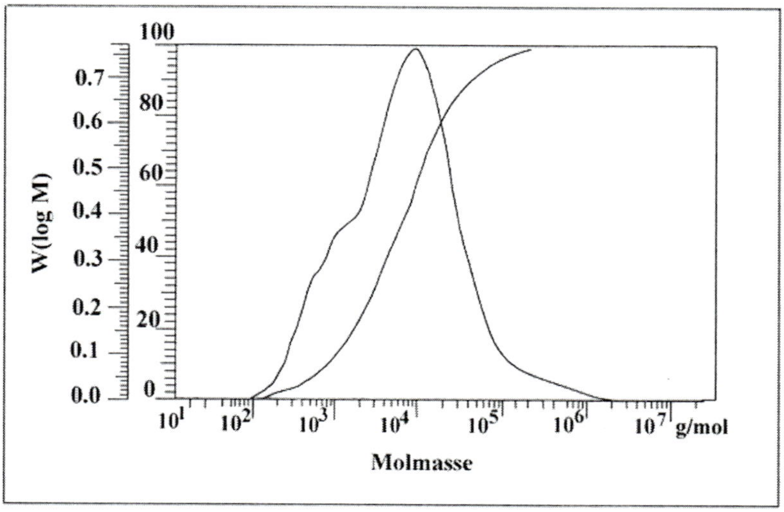

Figure 12. Gel permeation chromatographic curves of oligomer I.

Wide angle X-ray analysis showed that synthesized oligomers are amorphous one phase systems and value of interchain distance is equal to d_1=8,43 Å. As it is seen from the Table 1, the glass transition temperature of oligomers, depending of amount of cyclohexenyl groups in the side chain, changes in the range $-14^0C\div-10^0C$. Increasing glass transition temperature compared with linear polydime-thylsiloxane may be explained with the insertion of cyclohexenyl fragments in the chain and with the obtaining of partially branching systems.

Thermogravimetric investigations of oligomers have been carried out and 5% mass losses are observed at 230^0C. With the rise of temperature the mass losses increases and above 750^0C the changes of mass losses do not occur. With the insertion of cyclohexenyl fragments in the side chain the thermal-oxidative stability of oligomers increases.

By practical standpoint these above mentioned oligomers with unsaturated fragments in the chain are very interesting systems, because they contain unsaturated bonds, at the expense of the existed platinum chlorohydric acid in the system, gives possibility to change the properties of obtained sili-conorganic elastomers in a wide range.

ACKNOWLEDGMENT

This work was supported by DFG foundation. We are grateful to K. Mullen for help in this work.

REFERENCE

[1] Mukbaniani O.V.; Khananashvili L.M. *International Journal of Polymeric Materials*, 1994, *27*, 31.

[2] Aneli J.N.; Koberidze Kh.E.; Mukbaniani O.V.; Karchkhadze M.G.; Khananashvili L.M. *Russian Polymer News*, 1998, *3,* 7.

[3] Galashina N.M. *Vysokomolekuliarnye Soedinenye* (in Russian), 1994, *36A*, 640.

[4] Aneli J.N.; Mukbaniani O.V.; Karchkhadze M.G.; Koberidze Kh.E.; Tkeshelashvili R.Sh.; Khananashvili L.M. *Georgian Patent*, 1996, №1348101-96.

[5] Mukbaniani O.V.; Khananashvili L.M.; Esartia I.G.; Khaduri S.D. *International Journal of Polymeric Materials*, 1994, 24, 211.

[6] Mukbaniani O.; Khananashvili L.; and Zaikov G. *Aging of Polymers, Polymer Blends and Polymer Composites*, Nova Science Publishers, Inc., New York, 2002, 1, 99.

[7] ogaideli A.I.; Khananashvili L.M.; Nakaidze L.I.; Chogovadze T.V.; Tskhovrebashvili V.S.; Volkova R.V.; Nogaideli G.A. *Doklady Akademye Nauk GSSR* (in Russian), 1978, 92, 341.

[8] Mukbaniani O.; Scherf U.; Karchkhadze M.; Koberidze Kh. *International Journal of Polymeric Materials*, 2000, 48, 177.

[9] Dewar M.I.S.; Zeobish E.G.; Healy E.F.; Stewart J.S. *J. American Chemical Society*, 1985, 107, 3902.

In: Frontiers in Physical Organic Chemistry
Editor: G.E. Zaikov, pp. 41-57
ISBN 1-60021-128-3

Chapter 4

HYDROSILYLATION AND DEHYDROCONDENSATION REACTIONS OF METHYLHYDRIDESILOXANE TO ACRYLIC AND METHACRYLIC ACIDS

O. Mukbaniani[1,2,*], G. Zaikov[3], N. Pirckheliani[2], T. Tatrishvili[1], S. Meladze[1], Z. Pachulia[2] and M. Labartkava[1]
[1]I. Javakhishvili Tbilisi State University,
I. Chavchavadze Ave., 1, 0128 Tbilisi, Georgia
[2]Sukhumi State University, Branch of Tbilisi State University,
Djikia Str., 12, 0187,Tbilisi, Georgia
[3]Institute of Biochemical Physics Russian Academy of Sciences
4 Kosygin Street, Moscow 119991, Russia

ABSTRACT

The reaction of hydrosilylation of α,ω–bis(trimethylsiloxy)methylhydridesiloxane (n≈35) with acrylic acid at various ratio of initial compounds, in the presence of platinum hydrochloric acid has been investigated. It was established that in the presence of platinum hydrochloric acid competitive dehydrocondensation reactions take place with formation of oligomers with various link of chain. In high stages of reaction, three-dimensional systems were obtained.

By quantum-chemical half empiric AM1 method fully characterization in model reaction of hydride addition of methylhydridesiloxane to acrylic and methacrylic acid was carried out. For all initial, intermediate and final products, in model reaction of hydrosilylation of methyldimethoxysilane to acrylic and methacrylic acid, it was calculated heats of formations (ΔH_f) and change of enthalpy of model reactions ($\Delta\Delta H$) proceeding with various concurrent scheme. Starting from calorific effects of all possible direction of hydrosilylation reaction of methyldimethoxysilane to acrylic acid it was concluded that,

* *Corresponding author, E.mail: mukbaniani@ictsu.tsu.edu.ge

thermodynamically more profitable course of the reaction is the Farmer rule (1,2-addition).
The structures of synthesized oligomers were determined by IR and NMR spectra data. Gel permeation chromatographic, differential scanning calorimetric and X-ray analysis of synthesized oligomers were carried out.

Key words: dehydrocondensation, hydrosilylation, methylhydridesiloxane.

INTRODUCTION

In literature there are many information about hydrosilylation of acrylic acid [1-4]. By A.D. Petrov the reaction of hydride addition of triethylsilane to acrylic acid in the presence of Spier catalyst has been investigated and it was shown that the reaction proceed in "1-4" position, with formation intermediate product, by rearrangement of this product trimethylsilylated ether of propionic acid with 56% yield was obtained. The reaction proceeds according to the following scheme [1]:

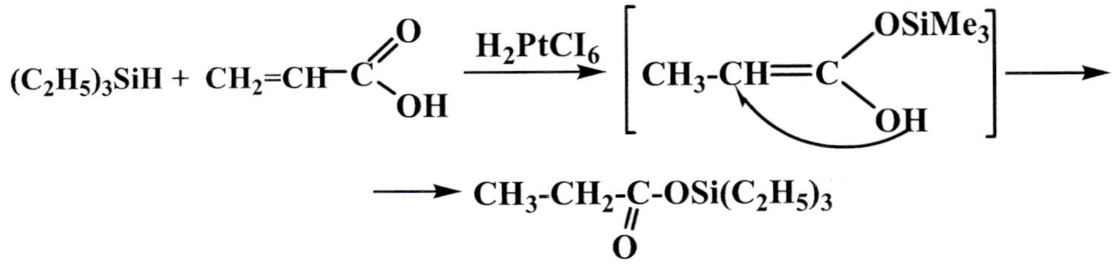

By the same author it was shown that the hydrosilylation acrylic methyl ether with triethylsilane proceeds in "1-4" position with formation unsaturated silylated methyl ether [2].

The reaction of hydride addition of α,ω–dihydridedimethylsiloxanes with acrylic and methacrylic acid has been investigated by K.A. Andrianov and his co-workers, in the presence of platinum hydrochloric acid and it was shown that hydrosilylation proceeds on an unsaturated C=C bond with Farmer rule receiving corresponding dicarboxylic acids. The reaction proceeds according to the following scheme [3,4]:

$$2CH_2{=}\underset{R}{\underset{|}{C}}HCOOH + H(SiMe_2O)_{n-1}SiMe_2H \xrightarrow{H_2PtCl_6} HOOC\text{-}CH_2CH_2(SiMe_2O)_{n-1}SiMe_2CH_2CH_2COOH$$

Where: R=H, Me; n=2÷11.

The structure of synthesized silicon containing dicarbon acids was determined by means of finding of molecular masses and IR spectra data.

The reaction of hydrosilylation of acrylic and methacrylic acid with 1,4-bis(dimethylsilyl)benzene in the presence of hydroquinone and Spier catalyst or platinum on the carbon was studied and it was shown that the reaction proceeds mainly by dehydrocondensation way, with formation of corres-ponding acyloxysilanes, according to the following scheme [5]:

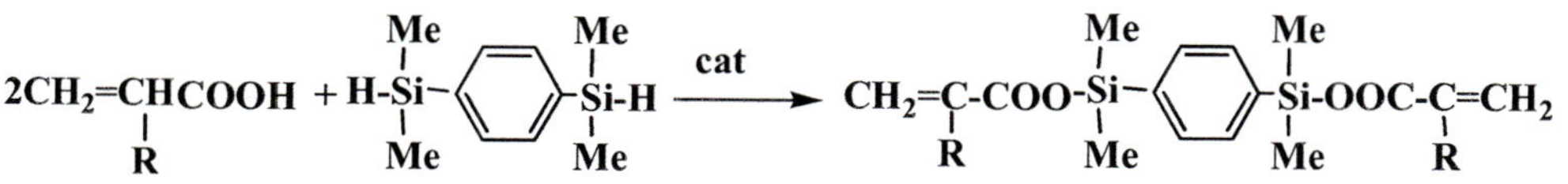

Where: R=H, Me.

In hydrosilylation reactions the hydroquinone was used for preventing the reactions of poly-merization process.

The reaction of hydrosilylation of methylhydridesiloxanes to unsaturated vinyl derivatives (alkenes, acrylic acid, acrylic ethers) in the presence of Spier catalyst has been investigated by the authors [6] and methylsiloxane oligomers with alkyl, carboxyl and ester side-groups has been obtained, accor-ding to the following scheme:

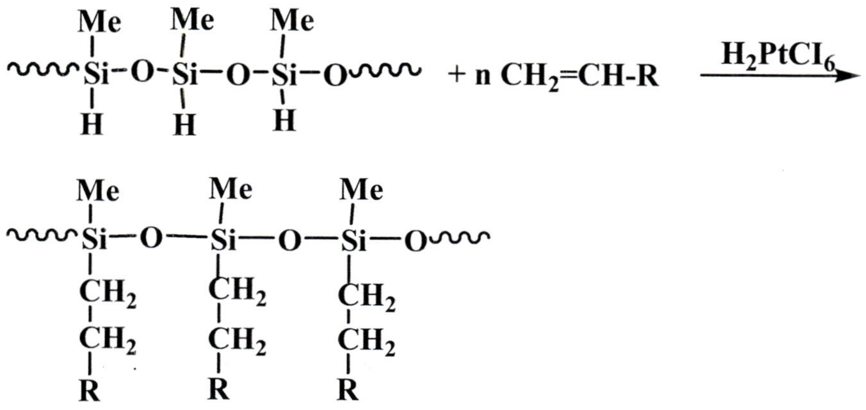

Where: R=C_nH_{2n+1}, -COOH, -COOAlk.

As against above-mentioned work [6] it was shown [7], that besides of hydrosilylation of methyl-hydridesiloxanes to acrylic acid in the presence of Platinum hydrochloric acid the reaction of de-hydrocondensation also takes place.

It is also known, that the reaction of catalytic dehydrocondensation between ≡Si-H containing com-pounds and hydroxyl containing organic compounds proceeds with isolation of hydrogen [8-10]. In the reactions of catalytic dehydrocondensation of hydrideorganosilanes with hydroxyl containing organic compounds are used such catalysts as colloid nickel, anhydrous zinc chloride, platinum hydrochloric acid [11,12]. It was shown [10] that in the reaction of dehydrocondensation of dihyd-rideorganocyclosiloxanes with dihydroxyorgano-cyclosiloxanes platinum hydrochloric acid works as caustic potassium and activation energy have the same values. On the other hand, it is known that platinum hydrochloric acid is catalyst of hydrosilylation [13]. The reaction capacity of ≡Si-H bond determined by catalyst nature and amount, reaction temperature, reactant concentration, solvent character, etc [8, 9]. By the authors [10] it was shown that during dehydrocondensation of hydride-containing organosiloxanes with hydroxyl-containing organosilanes and siloxanes in the presence of powder-like caustic potassium as a nucleophilic catalyst the competing reaction of dialkylsilo-xane elimination does not take place that is explained by a high rate of the reaction of dehydrocon-densation process.

EXPERIMENTAL PART

The starting materials for the synthesis of comb-type methylsiloxanes oligomers were used α,ω– bis(trimethylsiloxy)methylhydridesiloxane, acrylic and methacrylic acids.

The initial α,ω–bis(trimethylsiloxy)methylhydridesiloxane with the degree of polymerization n≈35, acrylic and methacrylic acids was received from Fluka. Before reaction, they acrylic and methacry-lic acids were distilled. The organic solvents used in reaction hydrosilylation were cleaned by dry-ing and distillation.

IR spectra all of samples have been taken on an UR-20 instrument from KBr pellets, while the ^{1}H NMR spectra on a "Perkin-Elmer" at operating frequency of 250 MHz. All spectra were obtained with the use of $CDCl_3$ as solvent and internal standard. A Perkin–Elmer DSC-2 differential scan-ning calorimeter was used to determine TGA and the thermal transitions (T_g) were read at the maxi-mum of the endothermic or exothermic peaks. Heating and cooling scanning rates were 10^0C/min.

Gel-permeation chromatography investigation was carried out with the use of Waters Model 6000A chromatograph with an R 401 differential refractometer detector. The column set comprised 10^3 and 10^4 Å Ultrastyragel columns. Sample concentration was approximately 3% by weight in toluene and typical injection volume for the siloxane was 5 μL. standardization of the GPC was accompli-shed by the use of styrene or polydimethylsiloxane standards with the known molecular weight.

Wide-angle X-ray diffractograms were taken on a "DRON-2" instrument (Russian model). A-CuK_α was measured without a filter, the angular velocity of the motor ω≈2 deg/min.

Dehydrocondensation Reaction of α,ω–Bis(Trimethylsiloxy)Methylhydridesiloxane with Acrylic Acid.

Catalytic dehydrocondensation reaction was carried out in a two-necked flask equipped with a tube for catalyst in-let, reflux condenser connected with a gasometer. Between the gasometer and re-flux condenser a cold trap and wash bottle were installed.

The reaction products were placed into the flask and thermostated in an oil bath until constant temperature was set up. Then the catalyst 0,1 M solution of platinum hydrochloric acid in tetra-hydrofuran ($5 \div 9x10^{-5}$ g per 1,0 g of starting substances) was introduced. After that, hydrogen start-ed release; hydrogen was collected in the gasometer. After the reaction has been finished 5 ml an-hydrous toluene was added, was filtered and the solvent was eliminated.

Hydride Addition of α,ω–Bis(Trimethylsiloxy)Methylhydridesiloxane to Acrylic Acid

The hyd-rosilylation reaction was carried out in a three-necked flask equipped with stirrer, reflux con-denser and calcium chloride drying tube. The initial reagents were placed in the flask and thermos-tated in an oil bath until constant temperature was achieved. 0,1 M solution of platinum hydroch-loric acid in tetrahydrofuran ($5 \div 9x10^{-5}$ g per 1,0 g of starting substance) was used as a catalyst. The reaction carried out in an argon atmosphere in anhydrous toluene solution, at temperature range 60^0C. Then 5 ml anhydrous toluene to the reaction mixture was added, filtered connected to the vacuum and the toluene and unreacted acrylic acid was removed at $30 \div 40^0$C.

The hydrosilylation reaction in the presence of platinum on the carbon carried out in the same manner.

RESULTS AND DISCUSION

As it is seen from the literary review, the reactions of hydride addition ≡Si-H bond containing com-pounds to acrylic and methacrylic acid proceeds on different directions and disagrees to each other. The presented paper deals with synthesis and studies of the properties of comb-type organosilo-xanes containing carboxyl groups in the side chain.

For the purpose of synthesis of organosiloxanes oligomers with carboxyl groups in the side chain the reaction of hydride addition of α,ω-bis(trimethylsiloxy)methylhydridesiloxanes ($n \approx 35$) to acry-lic and methacrylic acids at 1:35 ratio of initial compounds, in the presence of catalysts platinum hydrochloric acid or Pt/C (5%), at various temperature as in dry toluene solution (50-60%) so in melt condition has been investigated.

Preliminary heating of the initial compounds for 4 h in the presence of 0,1 M of platinum hydroch-loric acid solution in tetrahydrofuran ($\sim 5 \div 9x10^{-5}$ g per gram of reaction substances) in the tempera-ture range 50-70^0C showed that polymerization of acrylic or methacrylic acids, breaking of siloxane backbone, elimination of methane or hydrogen in methylhydridesiloxanes or other changes do not take place. Besides it was found that there are no changes in the NMR spectra of initial compounds. As it is known from literature [14] platinum hydrochloric acid except of the reactions of hydrosily-lation catalyzed the reaction of dehydrocondensation. Because of that, for completely characteriza-tion of the process during the reaction the conversion of hydrogen was observed.

The reaction of hydrosilylation investigated without solvent in the presence of catalysts at 70^0C temperature and it was shown, that at low stages of hydrogen conversion $\sim 10 \div 20\%$ the reaction proceeds with formation of three-dimensional systems. In this case, the depth of hydrosilylation reaction is about ~60%. For preventing the cross-linking processes, the reactions were conducted in ambience of dry toluene solution.

During the reaction in solution, it was established, that at ~35% hydrogen conversions (Figure 1) the cross-linkig processes take place. By determination of active ≡Si-H groups from soluble part of reaction products it was established, that the depth of hydrosilylation is about ~50%. This point that in the presence of platinum hydrochloric acid in dilute solution of toluene, besides of hydrosily-lation concurrent runs the reaction of dehydrocondensation ~30% (Figure 1, curve 1).

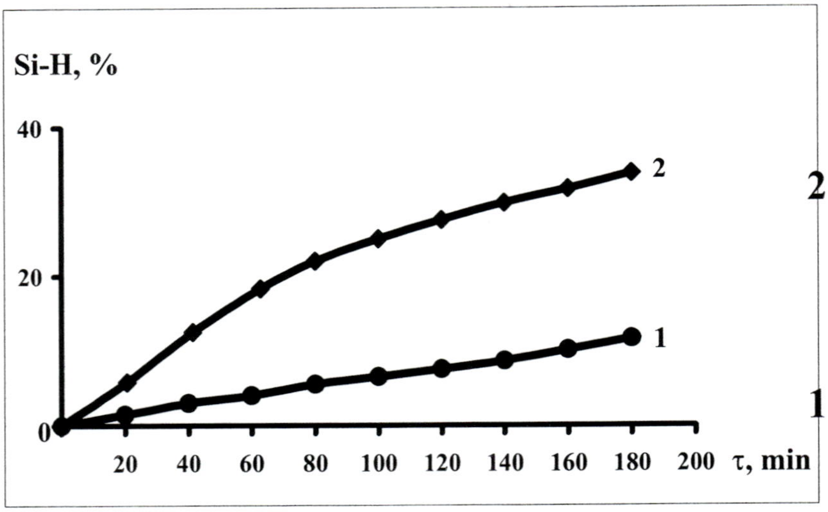

Figure 1. Dependence of changes of active ≡Si-H% groups on the time, upon of dehy-drocondensation reaction of α,ω–bis(trimethylsiloxy)methylhydridesiloxane to acrylic acid at 60^0C temperature. Where curve 1 corresponds to catalyst Pt/C, curve 2 corres-ponds to - H_2PtCI_6.

For the purpose of synthesis oligomers, well soluble in organic solvents the same reaction was studied in the presence of platinum on the carbon (5%) in dry toluene solution, but it was shown that really the reaction of dehydrocondensation proceeds less activity 9-10%, Figure 1 (curve 2), but the three-dimensional systems are obtained. Hydrosilylation in this condition proceeds about ~54%. Therefore, in both cases the reaction proceeds by two concurrent ways, with formation various- linked oligomers, according to the following scheme:

$$\equiv Si\text{-}H + CH_2{=}\underset{\substack{|\\ R}}{C}\text{-}COOH \xrightarrow{Cat} \equiv Si\text{-}C_2H_3R\text{-}COOH$$

$$\equiv Si\text{-}H + CH_2{=}\underset{\substack{|\\ R}}{C}\text{-}COOH \xrightarrow[-H_2]{Cat} \equiv Si\text{-}O\text{-}CO\text{-}\underset{\substack{|\\ R}}{C}{=}CH_2 .$$

With an increase of dilution of reaction solution the hydrosilylation and dehydrocondensation reaction rates decreases, but increases the soluble part of reaction products. Generally the reaction proceeds according to the following scheme:

$$Me_3SiO\left[SiMe(H)O\right]_m SiMe_3 + nHOOC\text{—}CH{=}CH_2 \xrightarrow{Cat}$$

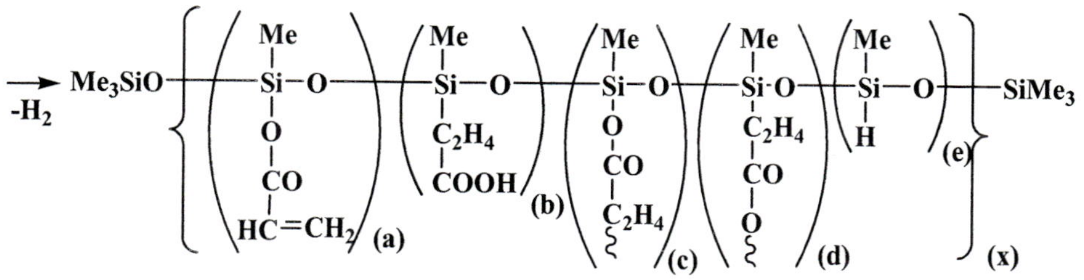

Where: m≈35; [(a)+(b)+(c)+(d)+(e)]x(x)=m. Cat – H_2PtCl_6, I (ratio 1:35); II (ratio 1:17,5) Pt/C – I^1 (ratio 1:35).

From cross-linked polymers the soluble part was isolated with the yield ~50-66%. They are color-less transparent viscous products soluble in ordinary organic aromatic type solvents. Some physical-chemical properties and the yields of synthesized branched oligomers are presented in Table 1.

Table 1. Some physical and chemical properties and the yields of branched oligomers

Oligomer, №	Yield, %	Ratio	η_{sp}^*	H_2 Conv., %	Depth of hydrosilylation, %	d_1, Å	$\bar{M}_\omega x10^{-3}$	T_g, ^{0}C
I	57	1:35	0,08	30	50	8,85	7,2	-67
II	49	1:17,5	0,06	23	49	-	-	-
I^1	66	1:35	0,08	10	54	8,23	-	-70
III	13	-	0,07	-	-	8,92	60,8	-23

*1% solution in Toluene at 25^{0}C.

It must denote that for a time the synthesized oligomers undergo to self-catalyzed structuring process. From oligomer I after 24 hours was isolated soluble part III with the yield ~13%.

For oligomer III, ^{1}H and ^{13}C NMR investigation carried out. As it is seen from Figure 2 one can ob-serve triplet signal, with chemical shift δ≈1,25 ppm (J=6,0 Hz) characteristic for hydrogen protons in ≡Si–CH_2- group. One can see also triplet signal for hydrogen protons in -CH_2-CO- group with chemical shift δ≈2,5 ppm and singlet signals with chemical shift δ≈0,1 and δ≈0,3 ppm for methyl protons in -$SiMe_3$ and ≡SiMe groups accordingly. In spectra one can see complicated multiplet characteristic for vinyl groups with chemical shift δ≈5,80÷6,1 ppm, which indicate that the reaction proceeds partially by dehydrocondensation way. There is low intensity singlet signal with δ≈4,2 ppm characteristic for unreacted ≡Si-H bond. One can see also, characteristic signal for residual acids protons with chemical shift δ≈10,2 ppm and multiplet signals of phenyl protons in toluene with chemical shift δ≈7,2÷7,6 ppm, which indicate the traces of toluene in synthesized oligomers.

In ^{13}C spectra of oligomer III (Figure III) it's well seen the signal for carbon in residual acid car-bonyl group with chemical shift δ≈179,3 ppm; signal with chemical shift δ≈7,5 ppm in ≡SiMe group, signal with chemical shift δ≈6,9 ppm in -$SiMe_3$; signal with chemical shift δ≈25,9 ppm characteristic for carbon in ≡Si–CH_2- group; signal with chemical shift δ≈27,5 ppm characteristic for carbon in methylene group in -CH_2-COOH fragment. Concerning of the reaction on the other direction, by dehydrocondensation way, is proved by presence in ^{13}C spectrum the signal with chemical shift δ≈171,7 ppm characteristic for carbon in free from proton carboxyl group ≡Si–O-CO-. Obtaining of cross-linked systems may be explained by intermolecular hydrosilylation and dehydrocondensation reactions:

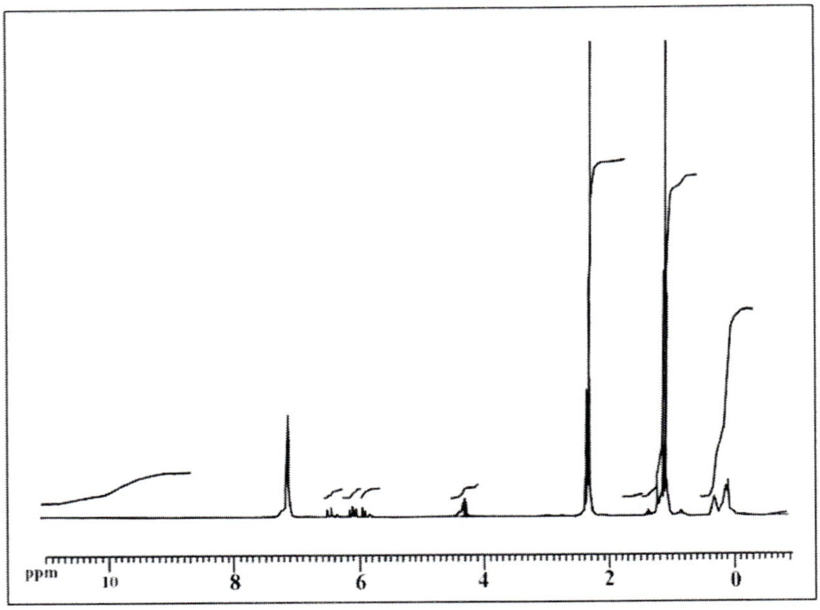

Figure 2. ^{1}H NMR spectra of oligomer III (solvent and standard $CDCl_3$).

$$-C_2H_4-COOH + H-Si\equiv \xrightarrow[-H_2]{Cat} \equiv Si-C_2H_4\text{-}CO\text{-}O-Si\equiv$$

$$\equiv Si-O-CO-CH{=}CH_2 \ + H-Si\equiv \xrightarrow{Cat} \equiv Si-O-CO\text{-}C_2H_4\text{-}Si\equiv$$

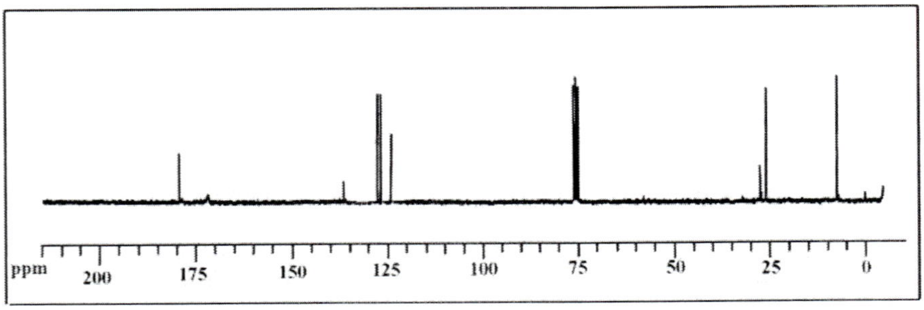

Figure 3. ^{13}C NMR spectra of oligomer III (solvent and standard $CDCl_3$).

In the IR spectra of oligomers one can see the absorption bands characteristic for ≡Si-O-Si≡ and ≡Si-O-C≡ bonds at 1020 and 1080 cm^{-1}. In the IR spectra one can see the absorption bands for ≡SiMe, for unreacted ≡Si-H bonds and intensive signal for carbonyl ethereal groups, signals for double bonds at 1275, 2165, 1720 and 1640 cm^{-1} accordingly.

It must be denote, that in IR spectra there is no absorption bands characteristic for associated and no associated hydroxyl groups. From which it is evident, that hydroxyl group take part in intermo-lecular dehydrocondensation as it is stated above, which led to formation of three-dimensional or cross-linked systems and in intramolecular cyclization according to the following scheme:

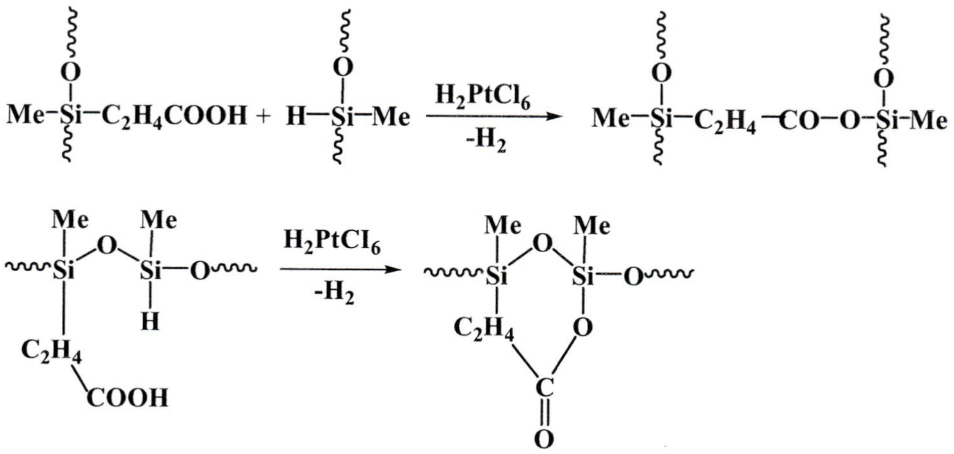

Where: $-C_2H_4- = -(CH_2)_2-$.

On the other side if the reaction proceeds by dehydrocondensation way, in this case in synthesized oligomers resonance signals for unsaturated vinyl fragments, with chemical shift in the range 5,5÷6,2 ppm, must be observed. But in the spectra Figure 2 and 3 there are low intensity resonance signals, which show that the fragment **≡Si-O-CO-CH=CH$_2$** may be entered in intra- and inter-molecular hydride addition by the scheme:

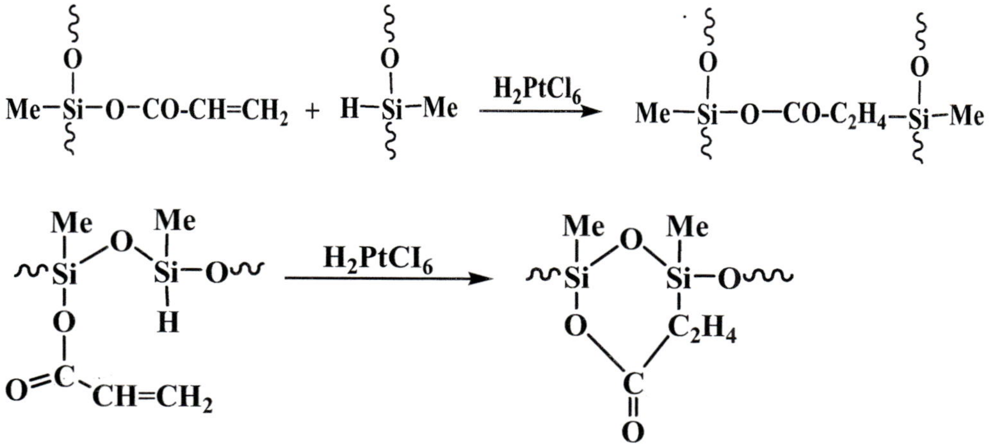

Where: $-C_2H_4- = -(CH_2)_2-$.

For oligomer I and III, wide angle X-ray investigation has been carried out. It was shown that oli-gomers are represented as one phase amorphous products, where the value of interchain distances changes in the range $d_1 \approx 8{,}23\text{-}8{,}92$ Å.

The synthesized oligomers were studied by GPC method. As it is seen from Figure 4 the GPC cur-ves of soluble part of oligomers I and III sharply distinguished from each other. Molecular weight distribution of oligomer I has bimodal character (Figure 4, curve 1) and for oligomer III polymodal character (Figure 4, curve 2), from which it's evident that during hydrosilylation reaction dehydro-condensation reactions takes place, that is in a contradiction with earlier published data [3,4]. Average weight molecular masses of oligomer I is equal to $\overline{M}_{\omega} \approx 7{,}2x10^3$ and polydispersity D≈9,4. For irregular branched oligomer III average weight molecular masses of oligomer III is equal to $\overline{M}_{\omega} \approx 60{,}8x10^3$ and polydispersity D≈35,3.

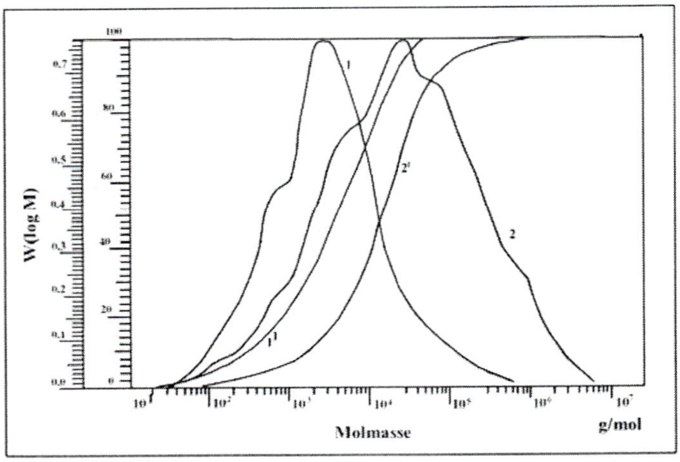

Figure 4. Gel-permeation chromatographic curves of oligomers. Where curve 1 corresponds to oligomer I and curve 2 corresponds to oligomer III.

Analogically hydrosilylation reaction of α,ω–bis(trimethylsiloxy)methylhydridesiloxane to meta-crylic acid in the presence of catalyst – platinum hydrochloric acid H_2PtCI_6 (0,1 M solution in tetrahydrofuran) at 1:35 ratio of initial compounds, at various temperature, both in a melt and in a dry toluene solution (50-60%) has been investigated.

It was shown that at 40^0C in case of 2% conversion of hydrogen the gelation processes take place. From soluble part of cross-linked products by determination of amount of active ≡Si-H group it was established, that the depth of hydrosilylation is about ~84%. With an increase of temperature from 50^0C and to 60^0C the part of the reaction of dehydrocondensation increases from ~3,8% up to ~15,0% (see Figure 5). The part of hydrosilylation reaction at 60^0C is equal to ~82%. So, during the reaction of α,ω–bis(trimethylsiloxy)methylhydridesiloxane to methacrylic acid in the presence of catalyst - H_2PtCI_6, besides of hydrosilylation reaction (both direction – Farmer rule and Markov-nikov rule), catalytic dehydrocondensation takes place.

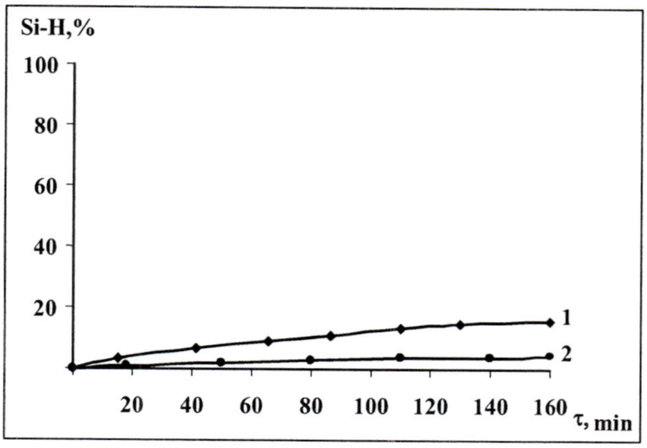

Figure 5. Dependence of changes of concentration active ≡Si-H% groups on the time, upon of dehydrocondensation reaction of α,ω–bis(trimethylsiloxy)methylhydridesiloxa-ne to methacrylic acid at 60^0C temperature. Where curve 1 corresponds to 50^0C, curve 2- 60^0C (in the presence of H_2PtCl_6).

During the dilution the depth of dehydrocondensation and hydrosilylation reactions decreases and the yield of soluble part rise. The reaction of methylhydridesiloxane with methacrylic acid proceeds with formation of various-linked products, according to the following scheme:

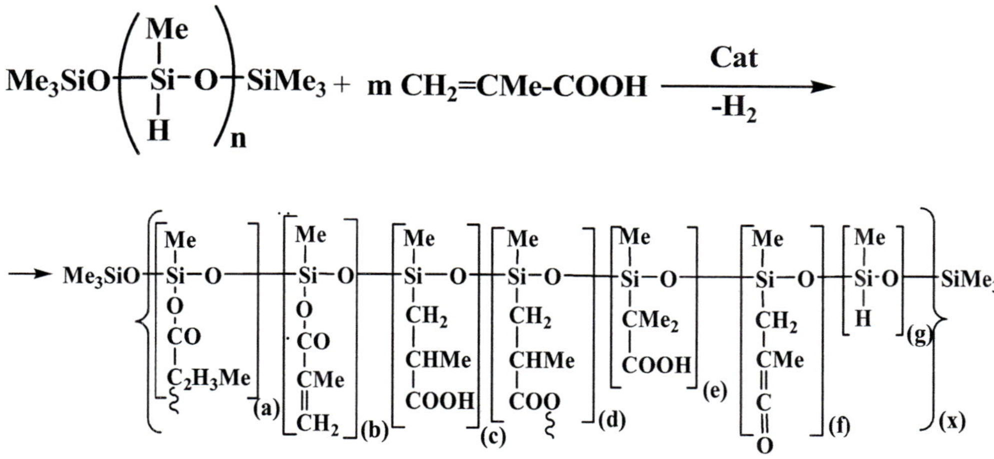

Where: [(a)+(b)+(c)+(d)+(e)+(f)+(g)](x)=n; 40^0C - IV^1; 50^0C - IV^2 and 60^0C - IV^3.

From cross-linked polymers, obtained branched oligomers are viscous products well soluble in aromatic solvents, with η_{sp}≈0,04-0,06. Some physical-chemical properties and the yield of oligo-mers IV^1-IV^3 are presented in Table 2.

For synthesized oligomers where studied by ^{1}H and ^{13}C NMR spectral data. Based on spectral data of reaction products proposed, that in generally hydride addition of methylhydridesiloxane to acrylic and methacrylic acid may be proceeds by several ways, see scheme 1:

Table 2. Some physical and chemical properties and the yield of oligomers IV^1-IV^3

#	Yield, %	Ratio	$\eta^*_{sp.}$	Reaction t-re T^0C	Convers-ion of H_2, %	Hydrosilylation, %	d_1,Å
IV^1	68	1:35	0,04	40	2,0	84	-
IV^2	66	1:35	0,05	50	3,8	88	-
IV^3	64	1:35	0,06	60	15	82	9,07

*In 1% solution of toluene at 25^0C.

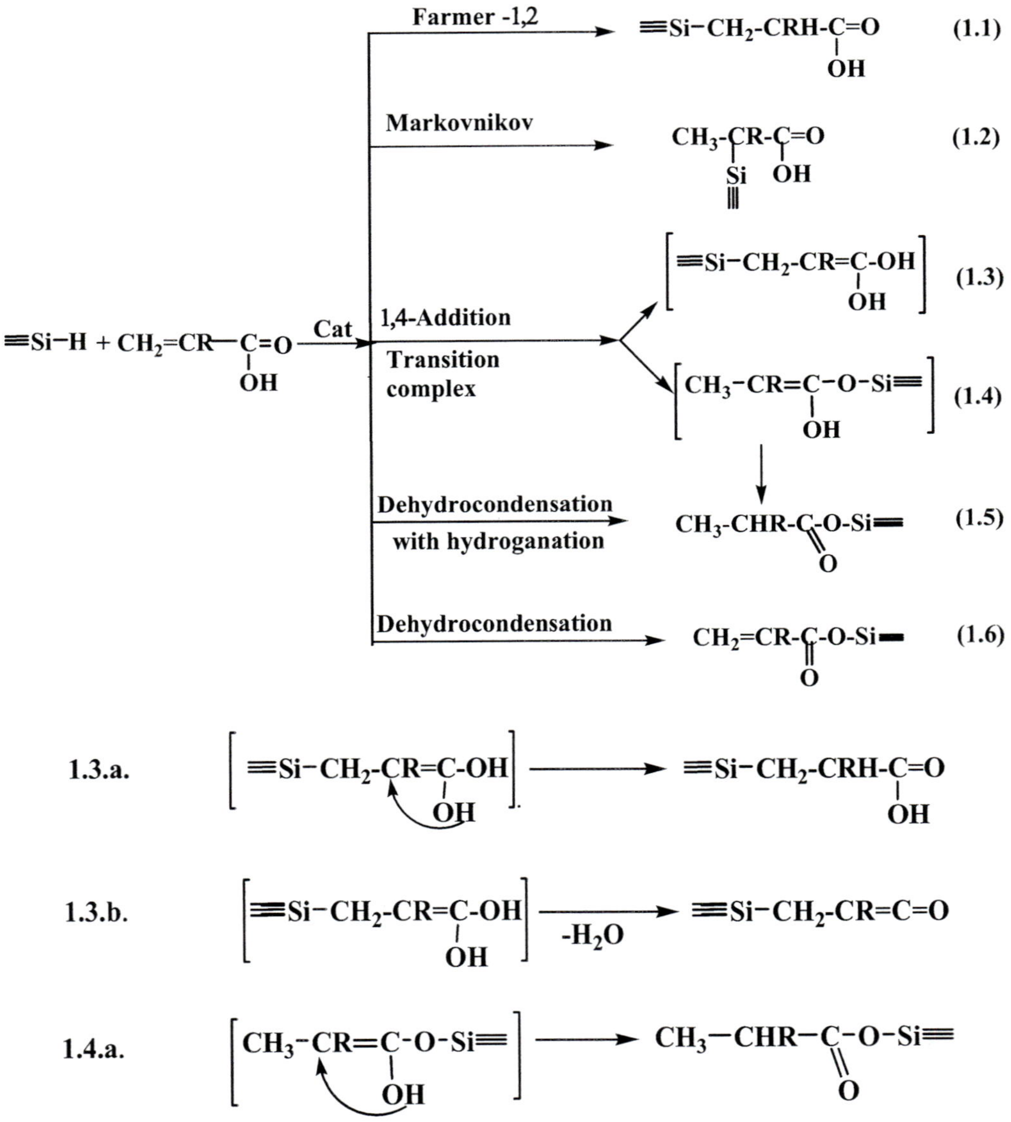

Scheme 1: The possible scheme of hydrosilylation of methylhydridesiloxane to acrylic and methacrylic acids

In ^1H NMR spectra of oligomer IV^3 one can observe singlet signal with chemical shift δ≈1,21 ppm for methyl protons in =CH-Me group, broadened singlet signal for protons of methyl group in C-Me group, signal with chemical shift δ≈1,22 ppm in α-addition product (Markovnikov rule, scheme 1.2), signal with chemical shift δ≈0,8 ppm corresponds to protons of methylene group, β-addition product in ≡Si-CH_2- fragment, which creation may be examine as structure obtained by regrouping of intermediate product (transition complex - 1,4-addition according to the Eltekov rule (scheme 1.3a).

The structure of the product obtained by above-mentioned scheme 1.3a is similar to the structure of the product obtained by Farmer rule (1,2-addition). Chemical shifts and multiplicity of remaining protons are: for methine group (≡CH) sestet (in case R=Me) with δ≈2,6 ppm (J=6,83 Hz), for methyl group singlet at δ≈1,21 ppm, (J=7,07 Hz) and for protons in carboxyl group broadened signal with chemical shift at δ≈10-11 ppm. In case of another course of reaction, after which elimination of molecule of water under the schema 1.3 takes place, with formation ketene group, their chemical shifts are following: for protons in methylene ≡Si-CH_2- group singlet signal with chemical shift δ≈0,78 ppm, for methyl protons in –CMe=C=O fragment singlet signal with chemi-cal shift δ≈1,96 ppm. It must be denote, that the authors [1] suggested direction of the reaction with obtaining of transition complex according to the scheme 1.4 earlier. Obtained transition complex then undergoes regrouping according to the scheme 1.4a with obtaining simultaneous dehydro-condensation-hydrogenation product like to the scheme 1.5. However, processing the reaction to above-mentioned scheme is improbable, because released hydrogen during dehydroconden-sation reaction, is eliminated away from reaction mixture.

Obtaining three various hydrosilylation products with methacrylic acid proved by ^{13}C NMR spectra data. Each of adduct are characterized with following chemical shifts: for hydrosilylation product by Markovnikov rule (scheme 1.2) resonance signal for methyl protons at δ≈18,11 ppm, signal for quaternary carbon nucleus at δ≈34,11 ppm and signal for carbon nucleus in carboxyl group at δ≈172,74 ppm. In hydrosilylation product by Farmer rule (1,2-addition, scheme 1.1), one can obser-ve signal for carbon nucleus with chemical shift at δ≈18,98 ppm in ≡Si-CH_2- fragment, signal at δ≈29,58 ppm for carbon nucleus in =CH- group and signal for methyl group at δ≈22,92 ppm in Me-CH- fragment.

One can observe analogically value of signals for compound containing ketene groups. Signal with chemical shift δ≈18,98 ppm characteristic for carbon nucleus of methylene group (≡Si-CH_2-), signal at δ≈30,2 ppm for carbon in methyl group [-C(Me)=C=O], signal for tertiary carbon nucleus at δ≈ 135,6 ppm [-C(Me)=C=O] and signal at δ≈183,43 ppm for carbon nucleus in the group =C=O.

As it is seen from ^{1}H and ^{13}C NMR spectra data hydrosilylation reaction of methylhydridesiloxanes to acrylic and methacrylic acid distinguished from each other and from literature data [1]. With ac-rylic acid hydrosilylation mainly proceeds by dehydrocondensation and hydrosilylation (Farmer rule (scheme 1.1 and 1.3a). In contrast of this, hydrosilylation to methacrylic acid proceeds almost in all direction presented on the scheme 1.

For fully characterization of hydride addition of methylhydridesiloxane to acrylic and methacrylic acid by quantum-chemical half empiric AM1 method [15], for all initial, intermediate and final products, in model reaction of hydrosilylation of methyldimethoxysilane to acrylic and methacrylic acid, it was calculated heats of formations (ΔH_f) and change of enthalpy of model reactions ($\Delta\Delta H$) proceeding with various concurrent scheme 2:

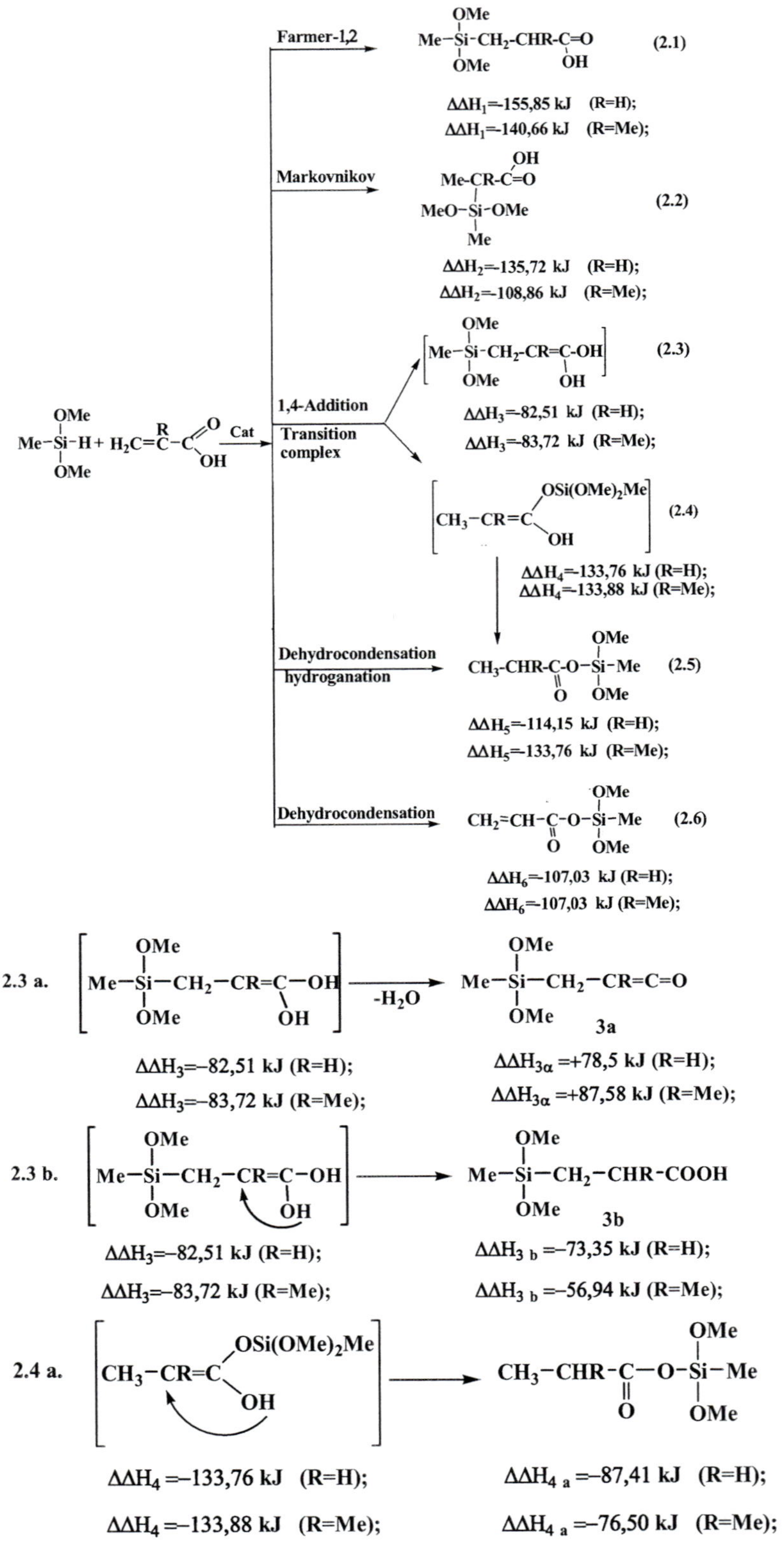

Scheme 2. Change of enthalpy of model reactions of hydrosilylation of methyldimethoxysilane to acrylic and methacrylic acid

As it is seen from the scheme 2, starting from calorific effects of all possible direction of hydro-silylation reaction of methyldimethoxysilane to acrylic acid thermodynamically is more profitable course of the reaction according to the Farmer rule (1,2-addition). It must be denoted that the pro-duct obtained by regrouping of intermediate product (1,4-addition, scheme 2.3) similar to Eltekov rule (scheme 2.3b), identical to product obtained by Farmer rule (scheme 2.1) and they have the same value of calorific effects [$\Delta\Delta H_1=\Delta\Delta H_3+\Delta\Delta H_{3a}$=-82,51 kJ+(-73,35 kJ)=-135,85 kJ]. Obtaining of ketene by regrouping of intermediate product according to the scheme 2.3a is most probable all.

As it is seen from ^1H and ^{13}C NMR spectral data of oligomer III, in the reaction of hydrosilylation of methylhydridesiloxanes to acrylic acids obtaining of ketenes is not observed.

The structure of the product obtained by dehydrocondensation-hydrogenation reactions (scheme 2.5) is identical to the product obtained by regrouping of intermediate product (scheme 2.4) accor-ding to the Eltekov rule. Possible scheme of the reaction in enlarged aspect may be represented as:

2.5a. 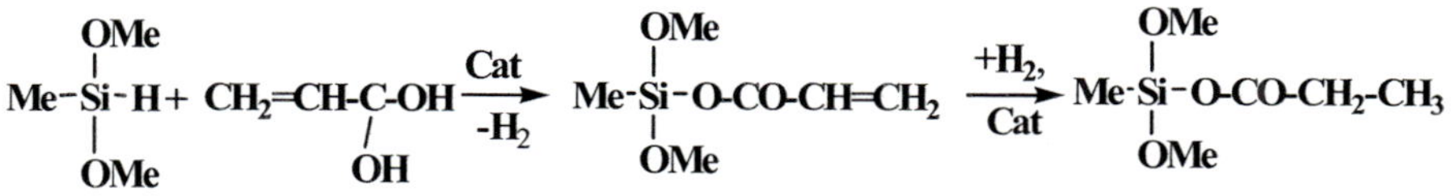

$\Delta\Delta H_6$=-107,02 kJ $\Delta\Delta H_5$=-114,15 kJ

The structure of the product obtained by this way is identical to the structure of the product obtained by regrouping of intermediate product (scheme 2.4) according to the scheme 2.5a and they have the same values of calorific effects ($\Delta\Delta H=\Delta\Delta H_4+\Delta\Delta H_{4a}=\Delta\Delta H_6+\Delta\Delta H_5$=-221,17 kJ). The course of the reaction according to the schemes 2.4 and 2.4a compared with another course energetically is pro-fitable, the dehydrocondensation-dehydrogenation course of the reaction according to the scheme 2.5a is improbable, because during dehydrocondensation reaction released hydrogen is isolated from reaction mixture, that is fixed from reaction mixture during the process, but strong denial of course of the reaction by the scheme 2.5a is impossible. Besides, it is possible that sensitivity of the instrument and conditions of received spectra was insufficient for identification of all possible pro-ducts obtained with a low concentration.

On the scheme 2 one can see also change of enthalpy of all reactions proceeding with various concurrent ways during the model hydrosilylation reaction of methyldimethoxysilane to methac-rylic acid.

On the assumption of heat of model reaction of methyldimethoxysilane with methacrylic acid (scheme 2), course of hydrosilylation according to Farmer rule (1,2-addition) thermodynamically is more profitable. The structure of the product, which is obtained by, regrouping of transition complex (1,4-addition, scheme 2.3) according to the Eltekov rule, scheme 2.3a, is similar to struc-ture of the product obtained by Farmer rule. Their heat of reaction {$\Delta\Delta H_1=\Delta\Delta H_3+$ +$\Delta\Delta H_{3b}$= [-83,72 kJ+(-56,94 kJ)]=-140,66 kJ} are same too. Formation of ketene group from transition complex ac-cording to the scheme 2.3a probable most of all, because its heat of reaction is equal to $\Delta\Delta H_{3a}$= +87,58 kJ. In spite of this, as distinct from acrylic acid, in methacrylic acid obtaining of ketene group is proved by ^{1}H and ^{13}C NMR spectra data.

With respect to dehydrocondensation-dehydrogenation reaction of methyl-dimethoxysilane with me-thacrylic acid which more comprehensive aspect represented as:

2.5a.
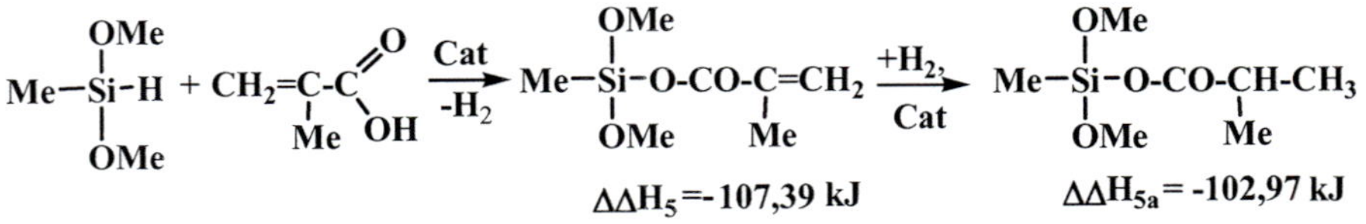

Analogically the course of the reaction according to above-mentioned scheme 2.5a is improbable, though its exclusion is not possible.

By wide-angle X-ray analyses of oligomer IV^3 it was shown, that oligomer is one phase amorphous system with value of interchain distances $d_1 \approx 9{,}07$ Å.

So, at the first time it was studied hydride addition reactions of methylhydridesiloxane to acrylic and methacrylic acid in the presence of catalysts platinum hydrochloric acid and Pt/C and differed from literature data it was shown, that besides hydrosilylation reactions (1,2- and 1,4-addition), de-hydrocondensation reactions takes place with formation various-linked thermo reactive oligomers.

References

[1] Petrov A.D., Sadikh-Zade S.I., Filatova E.I. *Zhurn. Obshch., Khimii*, 1959, v.29, p.2936.

[2] Sadikh-Zade S.I., Petrov A.D. *Zhurn., Obshch., Khimii, Chemistry and practical application siliconorganic compounds*, part 1, Leningrad, 1958, p.212.

[3] Andrianov K.A., Abkhazava I.I., Khananashvili L.M., Bagratishvili G.D. *Zhurn., Obshch. Khimii,* 1973, v.43, №7, p. 1505.

[4] Andrianov K.A., Abkhazava I.I., Khananashvili L.M., Bagratishvili G.D., Doksopulo T.P., Chagelishvili V.A. *Zhurn., Obshch. Khimii*, 1973, v.41, №1, p. 606.

[5] Gverdciteli I.M., Doksopulo T.P., Chikovani E.I. *Zhurn., Obshch. Khimii*, 1975, v.47, №7, p. 352.

[6] Nitzsche S. *Angew. Chem*., 1960, v. 72, №3, p. 112.

[7] Pirckheliani N.A, Meladze S.M, Chachua E.I., and Mukbaniani O.V. *Proceeding of the Georgian Academy of Sciences, Chem. Ser*., 2002, v.28, №1-2, p. 62.

[8] Dolgov B.N. *Chemistry and Practical Application of Organosilicon Compounds* Central Library of Technical Information, Leningrad, 1958, v.1, p.18

[9] Nogaideli A.I., Tkeshelashvili R.Sh., Nakaidze L.I., Mukbaniani O.V. *Works of Tbilisi State University*, 1976, v.167, p. 69.

[10] Mukbaniani O.V., Khananashvili L.M., Esartia I.G., Khaduri S.D. *Intern. Journ. of Polymeric Materials,* 1994, v.24, №1, p. 131.

[11] Dolgov B.N., Kharitonov I.P., Vorokvov M.G. *Zhurn., Obshch., Khimii*, 1954, v.24, p. 861.

[12] Baines J., and Eaborn C. *J. Chem. Soc*., 1955, p. 4023.

[13] Mukbaniani O.V., Scherf U., Gurgenidze G.N., Karchkhadze M.G., Meladze S.M., Kha-nanashvili L.M. *International Journal of Polymeric Materials*, 2001, v.48, №3, p. 267.

[14] Mukbaniani O.V., Gurgenidze G.N., Meladze S.M., and Khananashvili L.M. *Russian Polymer News,* 2001, v.6, №4, p. 18.

[15] Dewar M.I.S., Zeobish E.G., Healy E.F., Stewart J.S. *J. Am. Chem. Soc.*, 1985, v.107, p. 3902.

In: Frontiers in Physical Organic Chemistry
Editor: G.E. Zaikov, pp. 59-84

ISBN 1-60021-128-3

Chapter 5

ORGANOSILICON COPOLYMERS WITH CARBOCYCLOSILOXANE FRAGMENTS IN DIMETHYLSILOXANE BACKBONE

O. V. Mukbaniani, and G. E. Zaikov*
Sukhumi State University, 12, L.Djikia str., Tbilisi – 380087, Georgia
*N.M.Emanuel Institute of Biochemical Physics, Russian Academy of Sciences, 4 Kosygin st., Moscow 119991, Russia

ABSTRACT

Organosiloxane copolymers with carbocyclic fragments in the backbone can be synthesized by HFC reaction between primarily prepared bifunctional organocyclosiloxanes and organocyclocarbosiloxanes and α,ω–dichloro(dihydroxy)-dimethylsiloxanes. Another method of synthesis of copolymers with carbocyclic fragments in the backbone represents polymerization and copolymerization of organospirobicyclosiloxanes and polymerizing polycondensation of organospirobicyclo-silazasiloxanes with various organosilicon diols.
Discussed in this paper is synthesis of primary organosiloxane and carbosiloxane compounds with functional groups of organobicyclospirosiloxanes and organobicyclosilazasiloxanes. Methods of obtaining copolymers with carbocyclosiloxane fragments in the backbone are also discussed.

1. SYNTHESIS OF ORGANOSPIROBICYCLOSILOXANES AND ORGANOSPIROBICYCLOSILAZASILOXANES

To obtain organospirobicyclosiloxanes, at the first stage of the process initial 1,1-dichlorophenylcyclotri(tetra)siloxane was synthesized [1 – 3].

HFC reaction of 1,3-dihydroxytetraphenyldisiloxane and 1,5-dihydroxyhexa-phenyltrisiloxane with silicon tetrachloride has been studied. It is shown that the reaction process and structure of products depend on dilution and the ratio of reagents, but in the absence of hydrogen chloride acceptor, at most two chlorine atoms in a molecule of silicon tetrachloride are substituted. The reaction proceeds by the scheme as follows:

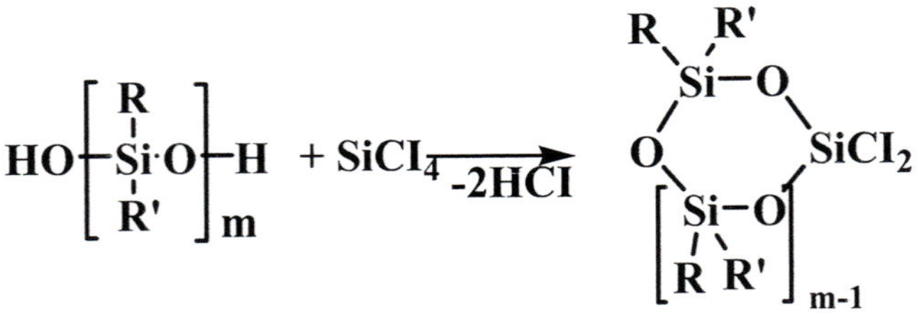

(5.1)

I

where R = R' = Me, Ph; R ≠ R', m = 2, 3.

It is found that affinity to homocondensation of dihydroxyorganosiloxanes increases with the size decrease of framing groups at the atom of silicon, which results in decrease of organochlorocyclosiloxanes content. Presence of hydrogen chloride acceptor in the reaction decreases the effect of diol side homocondensation and gives an opportunity to apply these reactions to direct synthesis of mentioned organocyclosiloxanes. However, in spite of the above-mentioned references, the study of HFC reaction between silicon tetrachloride and 1,3-dihydroxytetraphenyldisiloxane [4] displayed formation of bellied cyclohexasiloxane and its structural isomers in addition to dichlorotetraphenylcyclotrisiloxane. Table 1 shows some physical and chemical parameters of synthesized dichlororganocyclotri(tetra)siloxanes.

Table 1. Some physical and chemical properties of synthesized dichlororganocyclotri(tetra)siloxanes

N.	Compound	R	R'	*m*	Yield, %	T_{boil}, °C (*P*, mmHg)	T_{melt}, °C	Ref.
1	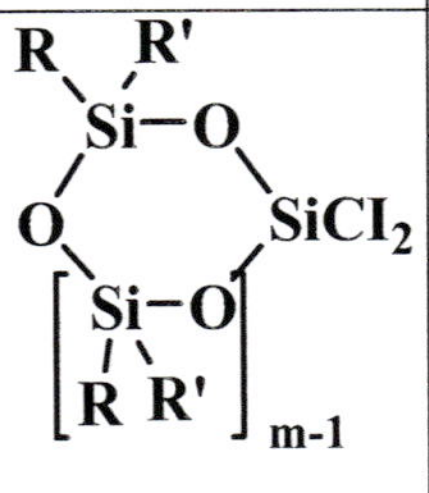	Ph	Ph	2	82	232 - 233 (1)	83-84	[3]
2		Me	Me	2	43	90 - 91 (750)	66-67	[3]
3		Me	tMe	3	67	81 - 82 (2)	30-31	[3]
4		Me	Ph	2	65	170 - 171 (2)	76-77	[2]

HFC reaction of 1,3-dihydroxytetraorganodisiloxane with silicon tetrachloride (the ratio of initial components 2 : 1 in the presence of acceptor, pyridine) proceeds with complete substitution of chlorine atoms and is accompanied by spirocyclization in accordance with the scheme as follows [5 - 8]:

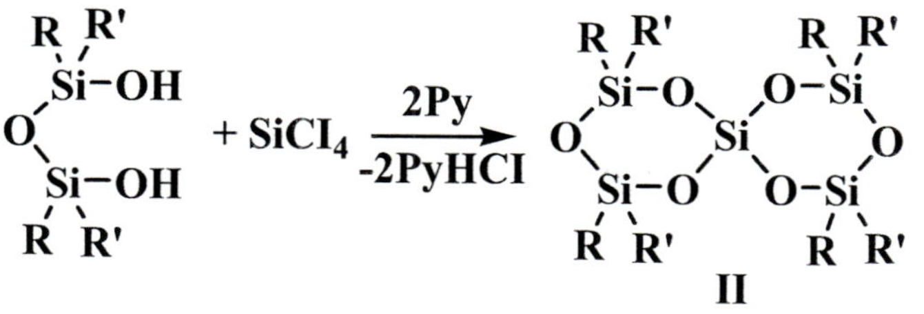
(5.2)

where R = R' = Me, Ph; R ≠ R'.

In the case of methyl framing, the reaction between meso-1,3-dimethyl-1,3-diphenyldisiloxanediol and silicon tetrachloride in the presence of pyridine gives one of stereo-isomers of 2,4,8,10-tetramethyl-2,4,8,10-tetraphenylspiro-(5,5)-pentasiloxane with the yield up to 35% in accordance with the following scheme [8, 9]:

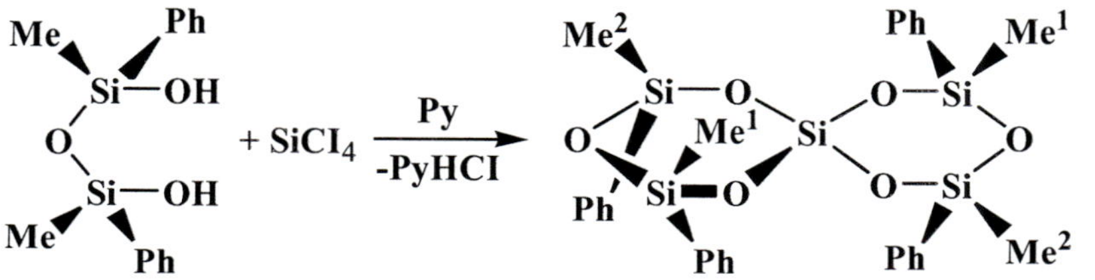
(5.3)

II

Spirobicyclosiloxanes were synthesized by HFC reaction of dichlorotetraorganocyclotrisiloxane with dihydroxydiorganosiloxanes (initial reagents ratio 1 : 1, in the presence of pyridine). The reaction proceeds as follows:

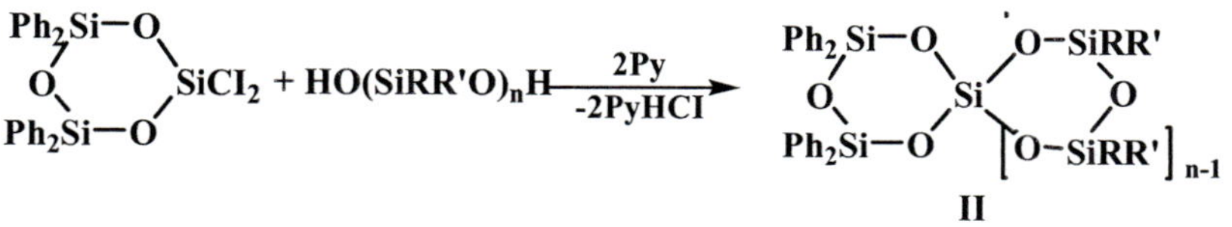
(5.4)

where R = R' = Me, Ph; R ≠ R'; $n = 2, 3$.

^{1}H ESR spectrum of the compound, obtained in reaction (7), where R = Me, R' = Ph, display responses from protons of methyl groups of both isomers. Chemical shifts of cis- and trans-isomer protons differ due to different contributions to screening, provided by circular currents of phenyl groups. In such systems [10] responses from protons of methyl groups, located in the cis-position to phenyl group. This allows relation of responses with $\delta \approx 0.44$ and $\delta \approx 0.53$ m.p chemical shifts to methyl protons of trans- and cis-isomers, respectively. Table 2 shows some physical and chemical parameters of synthesized spirobicyclosiloxanes.

For the purpose of synthesis of silazaoxane-siloxane and silazane-siloxane bicyclic compounds with spirane atoms of silicon the HFC reaction of 1,3-diamino-1,3-dimethyl-1,3diphenyldisiloxane and 1,3-diamino-1,3-dimethyl-1,3-diphenyldisilazane with dichloro-tetraphenylcyclotrisiloxane in diethyl ether (1 : 1 ratio of initial reagents) has been studied [11 – 13]. The reaction proceeds with formation of silazaoxane-siloxane (IIIa) and silazane-siloxane (IIIb) spirobicyclosiloxane:

Table 2. Some physical and chemical parameters of synthesized spirobicyclosiloxanes II

Compound, No.	R	R'	*n*	Yield, %	T_{boil}, °C (mm)	T_{melt}, °C	Chemical shift, δ(m.p.)	References
1	Me	Me	2	22	230-231 (0.5)	150-151	0.22	[7-9]
2	Me	Ph	2	48	265-266 (0.01)	114-115	Cis-trans 0.53; 0.44	[7-9]
3	Me	Ph	2	36	200-201 (0.01)	70	Cis 0.56; 0.48	[7-9]
4	Me	Ph	2	16	205-206 (0.01)	142-143	Trans- 0.42	[7-9]
5	Me	Me	3	51	214-215 (0.2)	54-55	0.05; 0.107	[7-9]
6	Ph	Ph	2	65	265-268 (0.01)	252-253	-	[1]
7*	-	-	-	36 16	200-201 (0.01) 205-206 (0.01)	70-71 142-143	Cis 0.56; 0.48 trans 0.42	[8]

Note: For compound 7, synthesized by reaction 3.

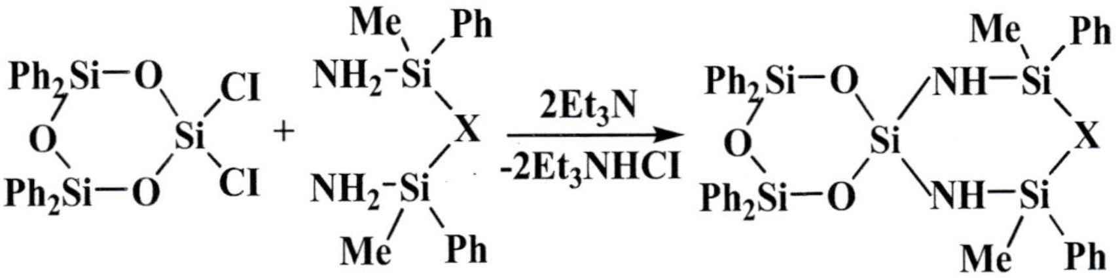

(5.5)

IIIa-IIIb

where X = O (IIIa), NH (IIIb).

The composition and structure of synthesized spirocyclosiloxanes are detected by the ultimate analysis, determination of the molecular mass, IR and ^{1}H ESR spectroscopy. Table 3 shows some physical and chemical parameters of silazaoxane-siloxane (IIIa) and silazane-siloxane (IIIb) spirobicyclosiloxane.

Table 3. Some physical and chemical parameters of silazaoxane-siloxane and silazane-siloxane spirobicyclosiloxane

No.	Compound	X	Yield, % (total)	Stereo-form	T_{boil}, °C (*P*, mmHg)	T_{boil}, °C	δ_{Me}, m.p.	Ref.
1	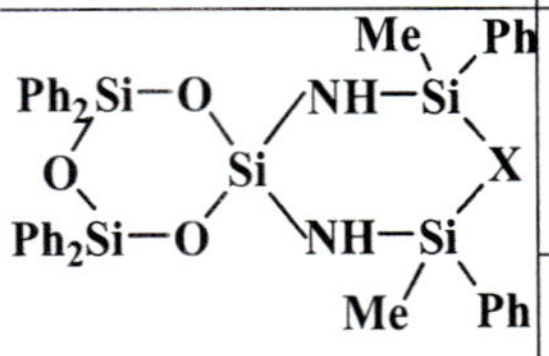	O	28	cis trans	265-270 (0.02)	- 134-135	0.475 0.350	[11]
2	**IIIa-IIIb**	NH	48	cis trans	299-301 (0.02)	170-171 -	0.475 0.340	[11]

ESR spectral analysis of spirobicyclic compounds has detected two existing isomers. ^{1}H ESR spectra of the compound with IIIa and IIIb structures display two responses of methyl protons, corresponded to cis- and trans-isomers.

It is known that for such systems [10] responses of protons in cis-oriented methyl groups in relation to phenyl ones are shifted towards stronger field, compared with responses of trans-methyl group protons. This occurs as a consequence of different contributions into shielding, stipulated by circular currents of phenyl groups at different mutual orientations of phenyl and methyl groups. This fact gave an opportunity to correspond the responses with chemical shifts $\delta \approx 0.475$ and $\delta \approx 0.475$ m.p. to cis-isomers, and the responses $\delta \approx 0.350$ and $\delta \approx 0.340$ m.p. to trans-isomers of compounds 1 and 2 (structures IIIa and IIIb, Table 4). It is of interest that spectra of compounds 1 and 2 (Table 4) possess equal chemical shifts of methyl protons responses that indicates that shielding of methyl groups in these cycles depends their orientation in relation phenyl groups only.

2. SYNTHESIS OF BIFUNCTIONAL HETEROCYCLIC COMPOUNDS

Bifunctional heterocyclic compounds were synthesized in two stages. At the first stage, the initial compound 1,1,1,3-tetrachloro-3,3-dimethyl-1,3-disilaethylene was synthesized. Reactions of 1,1,1,3-tetrachloro-3,3-dimethyl-1,3-disilaethylene synthesis by hydride addition of dimethylchlorosilane to vinyltrichlorosilane in the presence of catalyst, platinum-hydrochloric acid [14, 15], and dicobalt octacarbonyl [16] are known, which give high yield of the mentioned product. At the second stage, HFC reaction between 1,1,1,3-tetrachloro-3,3-dimethyl-1,3-disilaethylene and dihydroxydiphenylsilane in the presence of pyridine at different temperatures and 1 : 1 ratio of initial reagents proceeds in accordance with the scheme as follows [16]:

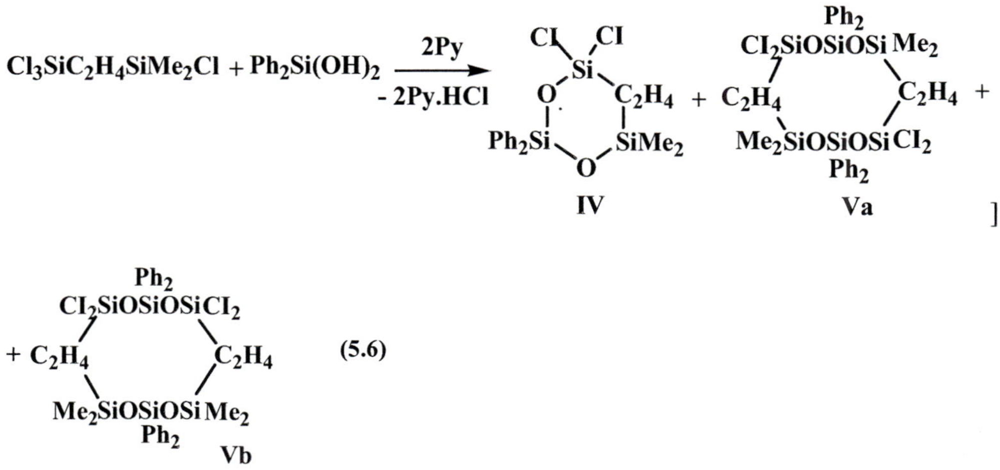

(5.6)

It is found that the yield of fractionated cyclic products is seriously dependent on the reaction temperature. HFC reaction proceeding at 20 - 25°C gives the yield of fractionated cyclic products below 70%, which at -5 ÷ -10°C is increased to 85%. Composition and structure of synthesized compounds were determined with the help of functional and ultimate analysis, determination of the molecular mass and from spectral data. Table 4 shows some physical and chemical parameters of carbocyclosiloxanes with functional groups.

Table 4. Some physical and chemical parameters of carbocyclosiloxanes with functional groups

No.	Compound	Yield*, %	T_{boil}, °C (P, mmHg)	T_{melt}, °C	References
1	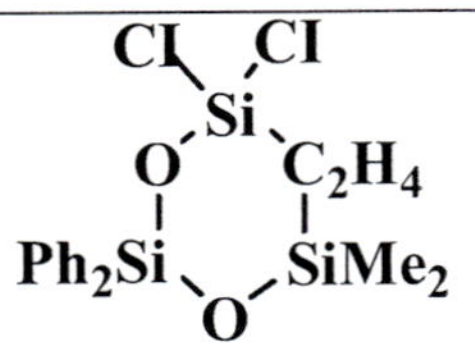	55 (68)	187-193 (0.01)	77-78	[15]
2	$[Cl_2SiC_2H_4SiMe_2OSiPh_2O]_2$	22 (17)	298-302 (0.01)	97-99	[15]

Note: * Yields at decreased temperature (-5 ÷ -10°C).

^{1}H ESR spectra of dichlorocyclocarbosiloxane with structure IV displays a complex multiplet for protons of asymmetrical methylene groups possessing chemical shifts in the range $\delta \approx 1.10 - 1.45$ m.p. with the center at $\delta \approx 1.30$ m.p. (4H), a singlet for methyl protons with chemical shift at $\delta \approx 0.45$ m.p. (6H) and a complex multiplet with chemical shift in the range of 7.32 - 7.75 m.p. ^{1}H ESR spectrum of a mixture of compounds Va and Vb is more complicated, because extended cycles may also contain other isomeric structures displaying different disposition of functional groups in a cyclic fragment, which cannot be rectified.

Gas-phase pyrolytic method [17, 18] is successfully applied to synthesis of silicon-containing heterocyclic compounds. Intramolecular gas-phase pyrolysis of o-dichlorosiloxydiphenyl was performed [17], which was accompanied by hydrogen evolution and formation of a six-membered cycle of the phenanthrene type. The yield of synthesized 10,10-dichloro-10-sila-9-oxarophenanthrene[1] equals 60%. The reaction proceeds as follows:

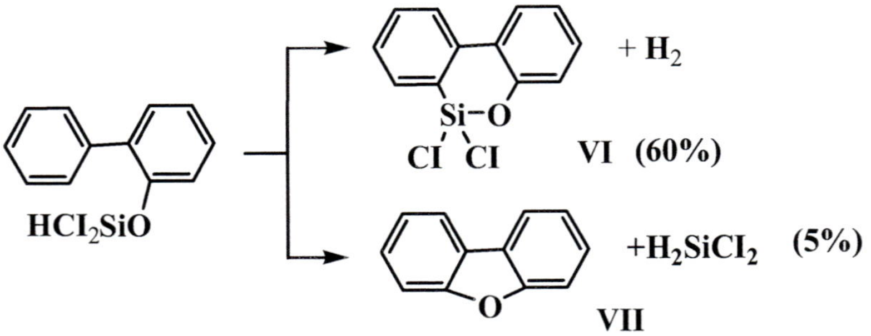

(5.7)

The above scheme indicates that the reaction partly proceeds with dibenz(o)furan formation. These investigators have also found that cyclization of *orhto*-siloxydiphenols can proceed not exclusively by ≡Si-H bond, but also by other bonds at silicon. However, this significantly decreases the yield of heterocycles. For *orhto*-trichlorosiloxydiphenyl, the yield of dichlorosilaoxarophenanthrene equals 5% only. The structure of

[1] Hereinafter, 1.1-dichloro-1-sila-2-oxadihydrophenanthrene.

dichlorosilaoxarophenanthrene was determined by physicochemical analysis, functional and ultimate analysis, from molecular mass and molecular refraction. It was also found that dichlorosilaoxarophenanthrene is quantitatively, without breaking ≡Si-O- bond, alkylated and arylated by Grignard reagents, and reduced by lithium aluminum hydride. When treated by hydrofluoric acid, it is transformed to difluorosilaoxarophenanthrene. High chemical stability of siloxarophenanthrenes indicates a definite aromaticity of these compounds (similar to aromaticity of boroxaro-borozarophenanthrenes [19, 20]. In accordance with the Dewar classification, the name of siloxarophenanthrene [19] was given to this heterocycle due to similarity of properties of the mentioned structures and possible aromatic type.

Hydrolysis of 1,1-dichloro-1-sila-2-oxadihydrophenanthrene in neutral medium in the presence of hydrogen chloride acceptor (aniline) has mainly produced corresponded dihydroxy derivative. The reaction proceeds as follows [21]:

(5.8)

VI VIII

After recrystallization of the reaction product from toluene-heptane mixture, 1,1-dihydroxy-1-disila-2-oxadihydrophenanthrene (VI) was obtained with 86% yield. After diluter evacuation from mother solution, dixydroxy-containing oligomeric product with structure VIII was obtained. Table 5 shows the yield and physical and chemical parameters of silaoxarophenanthrene with functional groups.

Table 5. Yield and some physical and chemical parameters of silaoxarophenanthrene with functional groups

No.	Compound	Yield, %	T_{boil}, °C (P, mmHg)	T_{melt}, °C	References
1	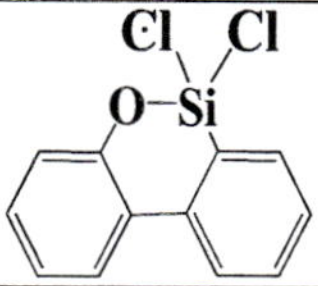	56	160 (1)	82-83	[17,18]
2	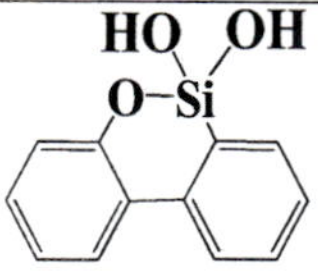	86	-	336-339	[21]

3. Synthesis and Study of Properties of Organosilicon Copolymers with Cardocyclosiloxane Fragments

Organosilicon copolymers with cardocyclosiloxane fragments in the backbone were synthesized by HFC reaction between 1,1-dichlorotetraorganocyclotrisiloxane and α,ω–dihydroxypoly(dimethylsiloxanes) [22]. Initially, condensation proceeds at room temperature, and at the final stage, for the purpose of obtaining rubbery polyorganosiloxanes, it is desirable to increase temperature to 100 - 150°C with simultaneous evacuation of diluter. Pyridine or diethylaniline were used as the acceptor of hydrogen chloride, produced during condensation. Generally, the reaction proceeds as follows:

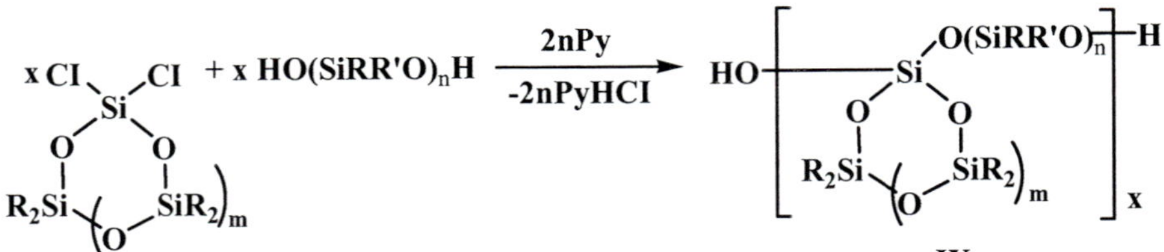

IX

(5.9)

where R = Me, Ph; n = 1 - 400.

As a consequence, organosilicon copolymers, which are transparent liquids with $[\eta] \approx 0.4 - 0.6$, highly soluble in organic solvents, with regular disposition of cardosiloxane fragments in the backbone were synthesized. Molecular masses of these compounds fall within the range of $1.2 - 1.9 \times 10^5$.

Cyclolinear poly(organosiloxanes) of the above-mentioned general formula represent liquids, capable of easy transformation to 3D-crosslinked elastomers at room temperature in the presence of catalytic amounts of acids and alkali (sulfuric acid, caustic alkali, ammonium and phosphonium bases), which induce regrouping of siloxane bonds. Gel formation of copolymers with structure XI in the presence of initiators, quaternary ammonium bases of the $Me_4N[SiMe_2O]_nNMe_4$ type, was studied [22]. It is shown that gel formation with elastomer networks proceeds during 15 minutes at room temperature in the presence of the above-mentioned initiator (0.02 - 0.1 wt.%). In this case, the gel-fraction concentration is 70 – 80%. By analogy, network systems are formed in the presence of sulfuric acid with 80% gel-fraction content. This property of the copolymers with cardocyclic fragments in the backbone makes them applicable as self-vulcanizing rubbers, filling compounds, and components for coatings and filler pastes.

At present, processes forming ladder-type polymers from organosilsesquioxanes are studied. In this connection, of great interest is catalytic polymerization of spirocyclic compounds. If organocyclosiloxanes are bifunctional compounds in polymerization and form linear macromolecules, organospirosiloxanes may display functionality two and four and, as a matter of principle, are capable of forming spatially crosslinked polymeric structures. It has been shown [23] that octamethylspiro-(5,5)-pentasiloxane polymerization does always give branched and, finally, crosslinked systems. It is also found [24] that polymerization of phenyl- and tolylspirobicyclosiloxanes in the presence of catalytic amounts of potassium naphthalene gives soluble polymers only. The authors think that polymerization proceeds with formation of polycyclic systems.

Polymerization of organospirobicyclosiloxanes with different organic groups in the framing was studied [25], specifically, polymerization of octaphenylspiro-(5,5)-pentasiloxane (compound 1, Table 6) and spirobicyclic compounds with different ratio of methyl and phenyl groups in the framing (cis-2,4-dimethyl-2,4,8,8,10,10-hexaphenylspiro-(5,5)-pentasiloxane, 2,2,4,4-tetramethyl-8,8,10,10-tetraphenylspiro-(5,5)-pentasiloxane and cis-2,4,8,10-tetramethyl-2,4,8,10-tetraphenylspiro-(5,5)-pentasiloxane – compounds 2 - 4, Table 6, respectively) in the presence of anionic polymerization catalysts.

Polymerization of cycles 2 and 3 (Table 6) was studied in benzene solution at 50°C in the presence of catalyst, $Me_4NO[SiMe_2O]_{26}NMe_4$. As indicated, it proceeds differently for compounds 2 and 3 in solution. Polymer synthesized from compound 3 is highly soluble and possesses specific viscosity of 1% solution in benzene equal 0.18; no gel formation was observed. IR-spectra indicate that about a half of six-membered cycles in the polymer are not disclosed, which, besides occurrence of a band at 1,090 cm^{-1}, is proved by occurrence of an intense doublet in the range of 1,025 - 1,040 cm^{-1} of the initial cycle spectrum and its preservation in the spectrum of reprecipitated polymer. Simultaneously, according to data of ESR spectra, conversion of cycles approaches 100%. This fact allows an assumption of linear or low-branched structure of the polymer containing six-membered cyclic fragments in the backbone:

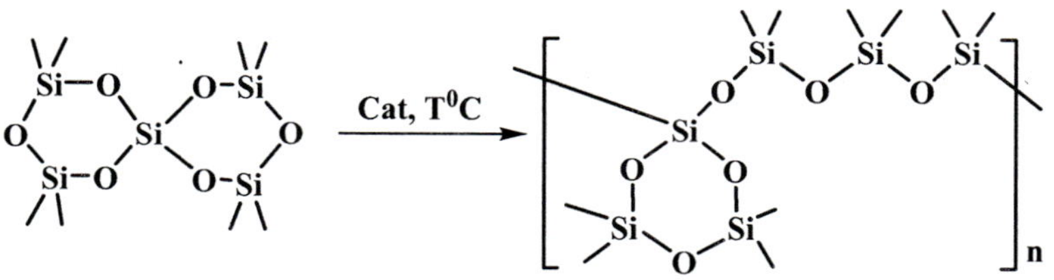

(5.10)

IX

Under the same conditions, polymerization of cycle 3 (Table 6) proceeds with formation of a transparent gel, the gel-fraction concentration in the reaction mixture being increased with time and approaching an equilibrium level. By analogy, compound 1 is polymerized at 100°C in the presence of ammonium catalyst in the case of preliminary solvent recovery; thus, the gel-fraction yield is somewhat higher in this case.

Polymerization of spirobicyclosiloxanes 1 – 4 (Table 6) in block at 150°C has given completely soluble polymer from cycle 1 only, which after reprecipitation possesses specific viscosity 0.1. For the rest spirobicyclosiloxanes, occurrence of gel-fraction accompanied formation of soluble polymers (Table 6). While phenyl group content in the molecule of the initial cycle decreased, the equilibrium yield of the gel-fraction increased.

Relatively equilibrium content of the gel-fraction in the polymer is the probability criterion for interchain interaction. That is why one may assume that the process of polymer synthesis proceeds in two directions, simultaneously: interchain interaction, which produces spatially crosslinked systems, and intramolecular polycyclization producing soluble polymers. If aromatic framing groups only are present in a spirocycle molecule, their intramolecular steric detachment promotes formation of soluble polycyclic products. While phenyl groups are substituted by methyl ones, the interchain interaction inducing gel formation becomes dominant.

Table 6. Gel-fraction yield in polymerization of spirobicyclosiloxanes at 150°C in block, in the presence of $KO(SiMe_2O)_8K$ as the catalyst – 0.02% (recalculated per KOH)

No.	Spirobicyclosiloxane	R	R'	R"	η_{spec}**	Gel-fraction yield, %
1	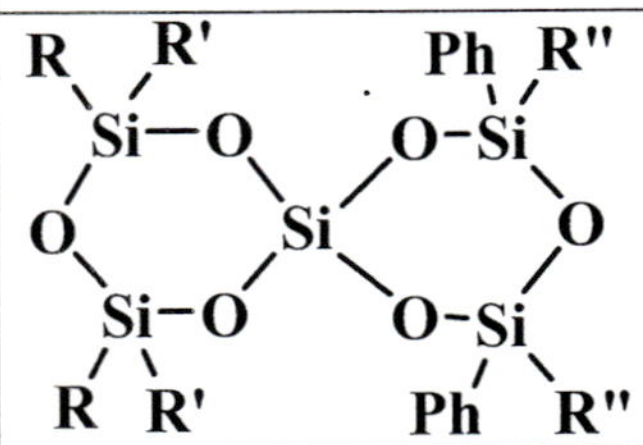	Ph	Ph	Ph	0.10	0
2		Me	Ph	Ph	0.09 (0.18)	28
3		Me	Me	Ph	-	56
4		Me	Ph	Me	-	64

Notes: * Copolymer viscosity after reprecipitation.
** 1% solution in benzene at 20°C.

Spirosiloxane 2 (Table 6), which under soft conditions produces a soluble polymer only, is capable of forming a crosslinked product with the polymerization temperature increase. Its polymerization in block at 120 - 150°C indicates that the equilibrium yield of the gel fraction is increased with the polymerization temperature.

Temperature, °C	120	130	150
Gel fraction yield, %	13	16	28
η_{spec} of soluble polymer	0.16	0.14	0.12

This is explained by the fact that temperature increase promotes an increase of mobility of active centers in propagating chains and probability of nucleophilic attack of cyclic fragments, not disclosed by them. Viscosity of the soluble part is slightly decreased with increase of polymerization temperature, which is probably associated with development of chain transfer reactions with a break. Proceeding of these reactions is proved by the results of ultimate analysis of obtained polymers.

Features of spirane and cyclic siloxanes' copolymerization was studied [26] on the example hexamethylcyclotrisiloxane (HMCTS) and 2,2,4,4-tetramethyl-8,8,10,10-tetraphenylspiro-(5,5)-pentasiloxane (compound 3, Table6) in benzene solution in the presence of tetramethylammoniumsiloxane. It may be expected that copolymerization is initiated simultaneously with the nucleophilic attack of both HMTCS and one of siloxane rings of spirocycle 3, because simultaneous disclosure of both rings of the spirocycle is of low probability. In this case, a linear poly(organosiloxane) with cyclic fragments, included in the siloxane chain, should be formed. As free HMCTS is exhausted, the future attack of these cyclic fragments must induce formation of statistically regulated polymer networks.

Figure 1a shows time dependencies of polymer specific viscosity for different ratio of HMTCS and compound 3 and constant total concentration of the catalyst, $[Cat]_{tot}$, which display an induction period in copolymerization, during which no viscosity increase in the system is observed. The induction period is increased with relative concentration of compound 3: at 8 mol% concentration no viscosity increase was observed during 140 minutes.

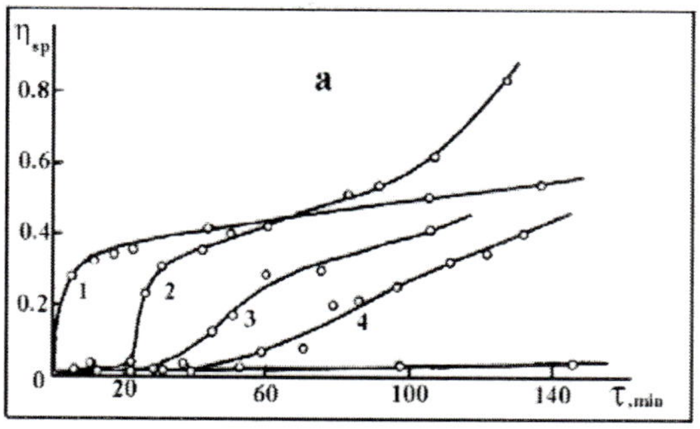

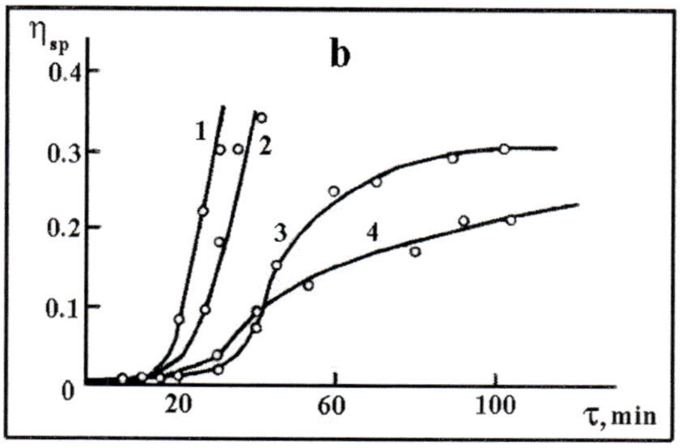

Figure 1. Specific viscosity variation during HMCTS and 2,2,4,4-tetramethyl-8,8,10,10-tetraphenylspiro-(5,5)-pentasiloxane (3) copolymerization: a – $[Cat]_{tot}$ = 0.015%; compound 3 concentration in the initial mixture, mol%: 1 – 0; 2 – 1; 3 – 2; 4 – 4; 5 – 8; b - $[Cat]_a$ = 0.5 wt.%; compound 3 concentration in the initial mixture, mol%: 1 – 8; 2 – 4; 3 – 2; 4 – 0.5.

After the end of the induction period, viscosity increases till insoluble product is formed. This happens, because raise of compound 3 concentration decreases the fracture of active centers of the catalyst falling at it ($[Cat]_a$ is the catalyst concentration in relation to compound 3 only). Thus it may be assumed that the induction period represents duration of compound 3 homopolymerization, caused by its much higher reactivity [27]. The rate of spirocycle restructuring, induced by the catalyst, is so higher, the higher is concentration of the latter. Hence, increasing is low, because compound 3 regrouping products apparently display the structure of globular polycyclic blocks. The viscosity increase, observed after complete depletion of spirocyclic compound 3 in the reaction mixture, happens at the expense of these blocks "overgrowth" by poly(dimethylsiloxane) chains with consequent formation of a polymer network due to the interchain exchange.

All the above-said is proved by the fact that at constant selected catalyst concentration in relation to compound 3 ([*Cat*]), the induction period is independent of the initial ratio between HMCTS and compound 3. For this series of experiments, the induction period equaled 16 minutes, approximately (Figure 1b). Vice versa, at constant initial HMCTS – compound 3 ratio (2 mol% of compound 3) the induction period is decreased with the increase of the catalyst concentration (Figures 2 and 3). As the catalyst concentration is $[Cat]_a$

= 2 wt.% (that corresponds to $[Cat]_{tot}$ = 0.11 wt.%) or higher, the induction period is not observed.

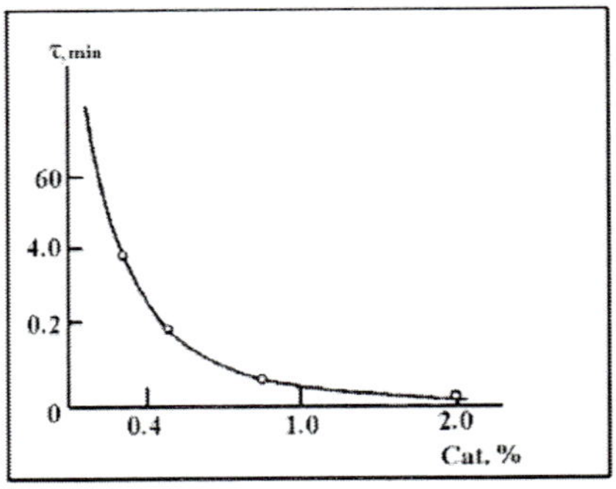

Figure 2. Dependence of the induction period on relative concentration of initiator, $[Cat]_a$, for HMCTS copolymerization with 2 wt.% of compound 3

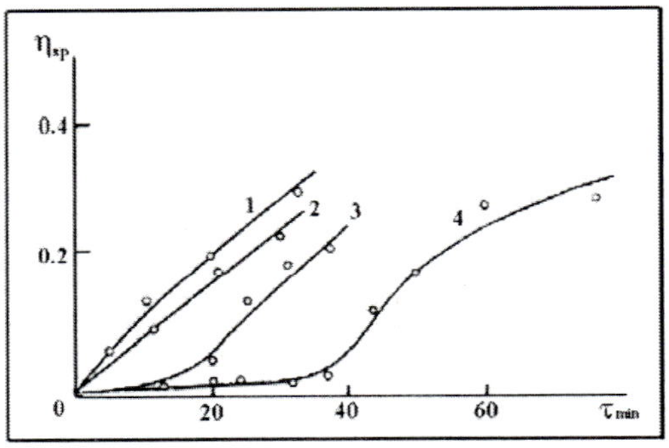

Figure 3. Variation of specific viscosity during HMCTS copolymerization with 2 wt.% of compound 3. $[Cat]_a$ ($[Cat]_{tot}$), wt.%: 1 – 2.0 (0.110); 2 – 1.0 (0.550); 3 – 0.50 (0.028); 4 – 0.28 (0.015)

Thus, experimental data give an opportunity to suggest that copolymerization of studied spirane and cyclic siloxanes is of the block-copolymerization type with polyspirocyclic blocks as the branching centers, but not via formation of linear-cyclic polymers with their consequent crosslinking.

Organosiloxane copolymers with cardocyclic fragments in the backbone, synthesized in the work [28], were used for development of artificial biocomposites, based on skeleton-free bone matrixes.

High strength (up to 140 MPa) and elasticity modulus (up to 20 GPa) of the bone tissue are stipulated by both hierarchic spiral structure of its protein and mineral parts and intensive interactions between them. The mineral part forms 70% of the bone composition.

To achieve an artificial biocomposite, natural protein, the reason for occurrence of bio-incompatibility, must be removed from the bone tissue with preservation of the mineral part structure, and porous bone matrix should be filled in with a bio-compatible material, able to polymerize.

Application of IR-spectroscopy and thermal gravimetry methods enabled to develop an optimal mode of bone treatment, which provides for complete elimination of collagen without breaking structure of the mineral part. After that the skeleton-free bone matrix was impregnated with siloxane oligomers in the presence of catalytic amounts of tetramethylammonium siloxane.

It is shown that the pore filling degree is highly dependent on molecular mass of the oligomer. The features of the polymerization process proceeding in pores and forming poly(organosiloxane) networks are studied. Network structures in pores of the bone matrix are formed due to splitting of cyclosiloxane fragments in the oligomer chain and their regrouping, the latter proceeding under the effect of tetramethylammonium siloxane and being not accompanied by occurrence of secondary low-molecular production. All components of synthesized composites are biocompatible and non-toxic for a man /human friendly/.

It is known from the literature [29 – 32] that organic and organosilicon diols interacting with silicon-nitrogen-containing heterocycles form polymers with rather high molecular mass and interesting physicochemical properties. To synthesize organosilicon copolymers with cardocyclosiloxane fragments in the linear chain, the reaction of 2,2,4,4,8,10-hexaphenyl-8,10-dimethylspiro-(5,5)-pentasila-7,11-diaza-1,3,5,9-tetraoxane (compound 1, Table 3) with different diols was studied [33].

Polycondensation was implemented in block or solution with cycle - diol molar ratio 1:1 till ammonia evolution is ended. Ammonia conversion and properties of produced substances are highly dependent on the chemical origin of initial diol (Table 7). In the case of α,ω–dihydroxypoly(dimethylsiloxane), the system became insoluble after 1.3 hour of condensation proceeding. Similar result was observed for diphenylolpropane both in block and in solution. A glassy-like low-melting polymer, soluble in benzene, is derived from 1,2-bis(hydroxymethyl)carborane. N,N'-bis(hydroxydimethylsilyl)tetramethylcyclodisilazane gives soluble oligomers; however, ammonia conversion in this case is low, both in block and in solution. The reaction proceeds as follows [33]:

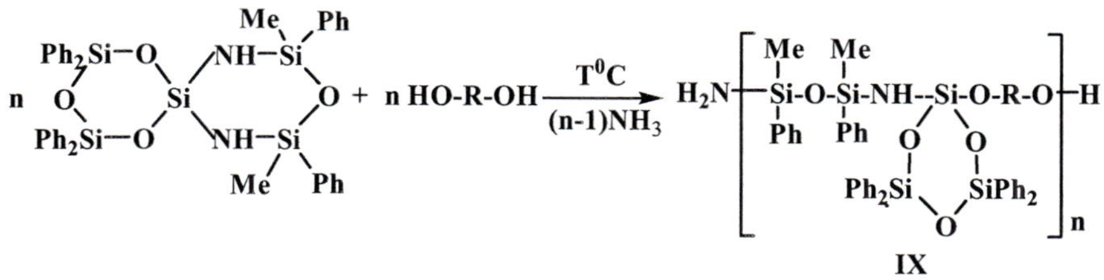

(5.11)

Table 7 shows conditions of synthesis for copolymers with cardosiloxane fragments in the backbone.

As shown above [31], condensation of organic and organosilicon diols with organocyclosilazasiloxanes and silazanes tends to form branched polymers and even polymer networks with nitrogen in the branching points. However, the formation mechanism of threesilyl-substituted nitrogen atom is not fully studied yet. Moreover, a break of the siloxane cycle may occur, and a branching center may be formed on the atom of silicon.

To clear up a possibility of realization of one alternative or another of branched polymer formation, model reactions have been performed, one of which between polydimethylsiloxane diols ($n \approx 25$ and $n = 2$) with hexamethylcyclotrisilazane (HMCTS) at 150°C in block. Ammonia liberation curves are shown in Figure 4. No formation of insoluble products was observed.

Table 7. Reaction conditions for 2,2,4,4,8,10-hexaphenyl-8,10-dimethylspiro-(5,5)-pentasila-7,11-diaza-1,3,5,9-tetraoxane with diols

No.	Diol	Medium	*T*, °C	Time, h	Quantity of liberated NH_3	Gel-fraction, %
1	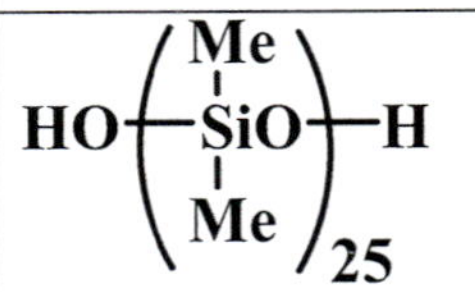	In block	180	5	70	66*
2	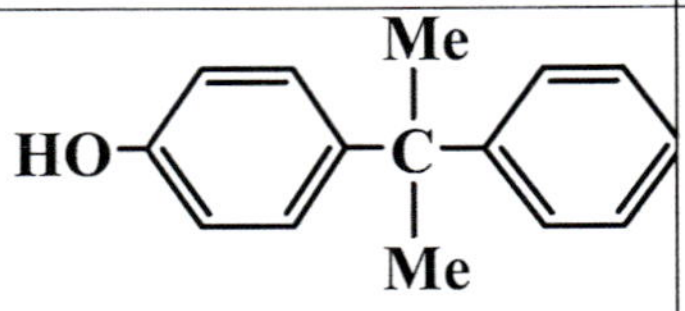	In block Toluene	150 110	6,5 3,5	86 90	73 31
3	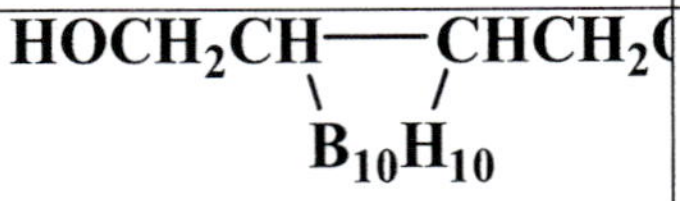	Toluene	110		99	_**
4	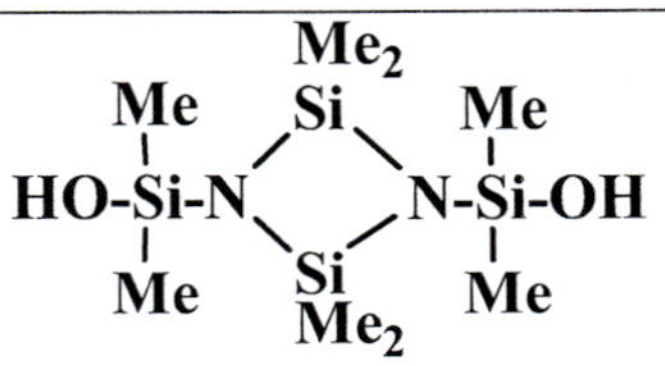	In block Toluene	140-200 110	4 9,5	48 40	- -

Notes: * 10 minutes before structuring, η_{spec} of the product in benzene equals 0.2.
** $\eta_{spec} = 0.1$.

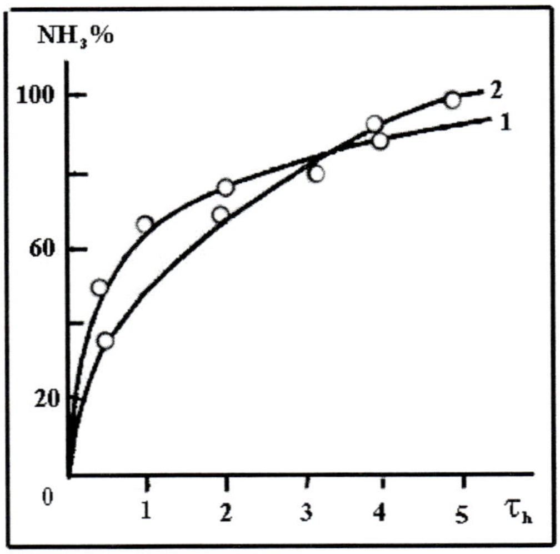

Figure 4. Ammonia liberation in the reaction between HMCTS and α,ω–dihydroxypoly(dimethylsiloxanes): $1 - n = 2$; $2 - n \approx 25$

Maximum viscosity of 1% polymer, derived from diol with $n \approx 25$, solution in benzene reached the value of 0.82, and for diol with $n = 2$ – 0.64. To clear up if the siloxane bond is broken in condensation of spirocyclic compounds with diols, HMCTS reaction with diol (n = 2) in the presence of hexamethylcyclotrisiloxane and the molar ratio of reagents 1:1:1 was performed. It was found that hexamethylcyclotrisiloxane is gradually eliminated from the reaction mixture (Figure 5.5), i.e. stressed siloxane trimer is broken under condensation conditions. Generally, the process may be described by a sequence of reactions as follows:

$$\equiv SiOH + HN(Si\equiv)_2 \longrightarrow (\equiv Si)_2O + H_2NSi\equiv \qquad (5.12)$$

$$\equiv SiOH + H_2N\,Si\equiv \longrightarrow (\equiv Si)_2O + NH_3 \qquad (5.13)$$

$$\equiv SiOH + NH_3 \rightleftharpoons \equiv Si\overset{-}{O}\overset{+}{N}H_4 \qquad (5.14)$$

$$\equiv \overset{*}{Si}\overset{-}{O}\overset{+}{N}H_4 + (\equiv Si)_2O \longrightarrow \equiv \overset{*}{Si}\text{-O-}Si\equiv + H_4\overset{+}{N}\overset{-}{O}Si\equiv \qquad (5.15)$$

Thus all three components, including hexamethylcyclotrisiloxane added to the active end of propagating chain by the last reaction, participate in the copolymer synthesized:

$$\equiv Si\overset{-}{O}\overset{+}{N}H_4 + HN(Si\equiv)_2 \rightleftharpoons \equiv SiOH + H_4\overset{+}{N}\overset{-}{N}(Si\equiv)_2 \qquad (5.16)$$

$$(\equiv Si)_2\overset{-}{N}\overset{+}{N}H_4 \longrightarrow \begin{cases} \xrightarrow{+(\equiv Si)_2O} (\equiv Si)_3N + H_4\overset{+}{N}\overset{-}{O}Si\equiv \\ \xrightarrow[+HN(Si\equiv)_2]{} (\equiv Si)_3N + H_4\overset{+}{N}\overset{-}{N}Si\equiv \end{cases} \qquad (5.17)$$

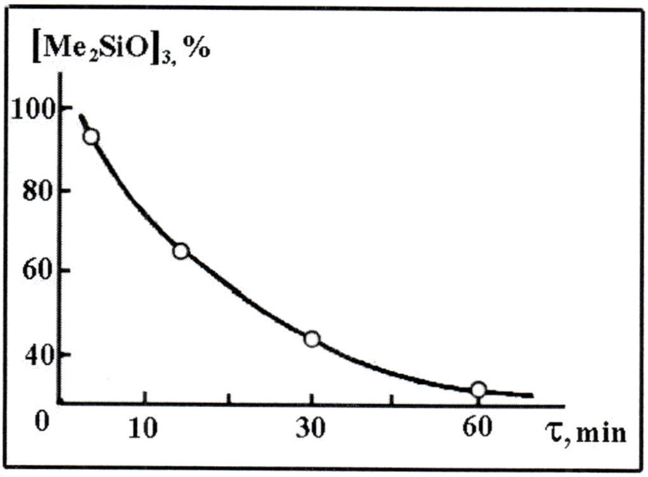

Figure 5. The exhaustion curve for $[Me_2SiO]_3$ in the reaction with HMCTS and diol ($n = 2$)

However, concentration of trisilylamino groups is assumably low, and the response of methyl protons, typical of these groups, was not observed in the ESRT-spectrum.

Transferred from model reactions to the organospirosilazasiloxane – diol system, it has been concluded that diol with lower acidity, for example, 1,2-bis-(hydroxymethyl)carborane gives a soluble product, whereas siloxane diols and diphenylolpropane form gels. In the case of carborane diol, the equilibrium reaction between it and ammonia is shifted to the left (see above) and the rate of siloxane cycle disclosure is negligible due to low concentration of nucleophilic particles, whereas disclosure of silazane cycle and condensation of OH- and NH-groups proceeds more completely (ammonia conversion approaches 100%), which leads to formation of linear or low-branched polymer. In the system with organosilicon and aromatic diols, concentration of nucleophilic particles capable of disclosing the siloxane cycle is higher. As a consequence, resulting comparatively rapid formation of branched polymers and polymer network and change in molecular mobility, the rates of polycondensation reactions and their completeness is lower than in the previous case.

Of special attention is N,N'-bis(hydroxydimethylsilyl)tetramethylcyclodisilazane. The reaction is started from HMCTS disclosure and condensation of NH_2- and OH-groups. However, introduction of ammonia into the system initiate anionic regrouping, proceeding at an appreciable rate as follows:

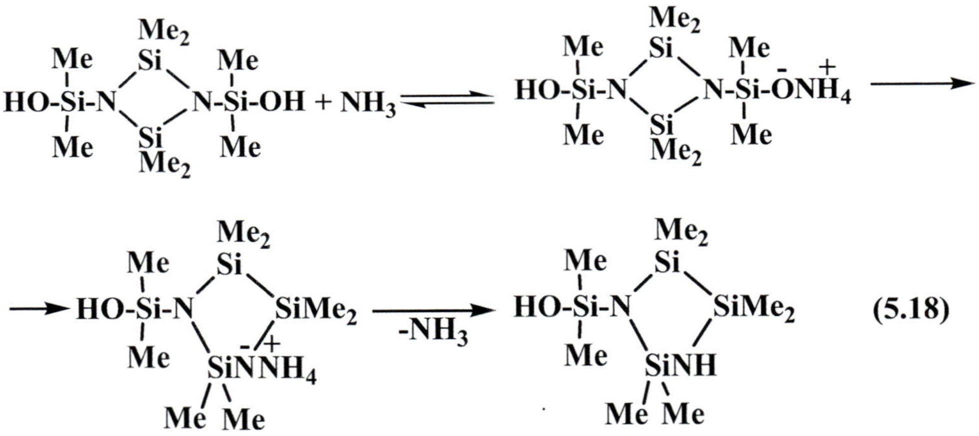

(5.18)

As a result, concentration of hydroxyl groups is decreased, polycondensation proceeds at lower rate and completeness, and low-molecular oligomers are synthesized. Copolymers obtained possess high resistance to thermal oxidation: weight loss is initiated at 350°C and the main degradation process proceeds at temperatures above 400°C.

Thus, application of polymerization of spirobicyclosiloxanes with cyclosiloxanes, as well as polymerizing polycondensation of organospirobicyclosilazasiloxanes with dihydroxy-derivatives may not always induce synthesis of organosiloxane card-type copolymers with the given structure of macromolecular backbone.

4. ORGANOSILICON COPOLYMERS WITH CARD-TYPE HETEROCYCLIC FRAGMENTS IN THE LINEAR CHAIN

Card-type cyclolinear copolymers containing cyclocarbosiloxane fragments and silaoxa-dihydrophenanthrene fragments and in the chain were synthesized by HFC of dichlorocyclo-carbosiloxane (structure IV, compounds 1, Table 4) [16,34,35] and 1,1-dichloro-1-sila-2-oxadihy-drophenanthrene (structure VI, compounds 1 and 2, Table 5) [36,37] with α,ω-dihydroxydime-thyl(diphenyl)siloxane with various length of siloxane chain, at 1:1 ratio of initial compounds in the presence of pyridine, as an acceptor of hydrogen chloride, was investigated. As its known, from literature the use of the acceptor amines leads to increasing of depth of HFC [38]. The reaction was carried out at a room temperature, at the completing stage the reaction mixture heated up to boiling point of used solvent.

The secondary reactions of homofunctional condesations of α,ω–dihydroxy-dimethylsiloxa-nes in such conditions does not take place. Earlier by the authors [39], it was shown that homofun-ctional condensation of 1,3-dihydroxytetramethyldisiloxane does not take place in the presence of more alkali amines, than pyridine.

It was shown, that at a small length of the linear dimethylsiloxane segment (n≤4) the copo-lymers yield is low, which was explained by the fact, that besides the intermolecular condensation, there partially takes place the intramolecular cyclization, with formation of cyclic compounds. These two ways of HFC during synthesis of cyclolinear copolymers were suggested early [40-44]. The reactions of HFC proceeds according to the following scheme:

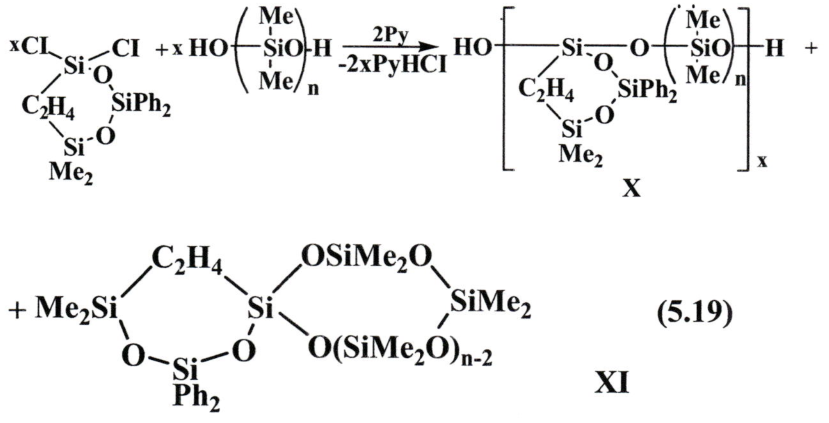

X

XI

(5.19)

where: n=2 – 36.

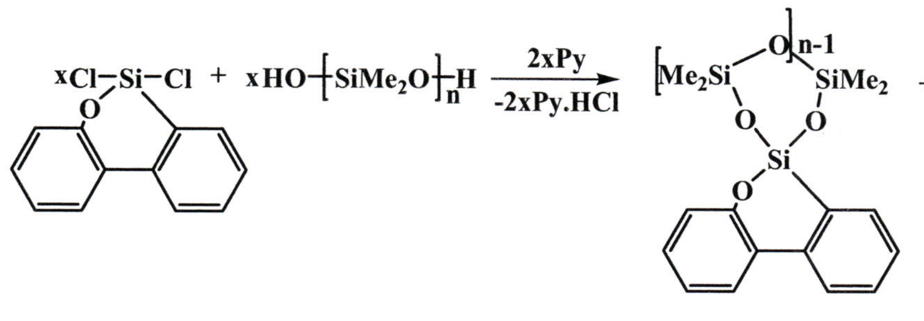

XII

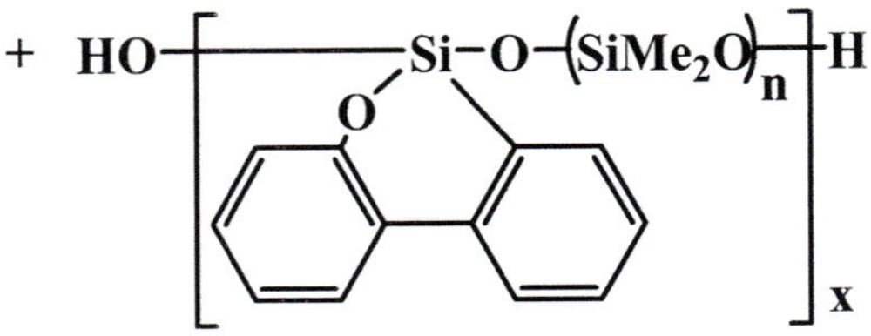

(5.20)

XIII

where: n=2 - 27.

The synthesized copolymers with structure X and XIII were reprecipitated from toluene solution by methyl alcohol. The yield of obtained copolymers is about 73÷95%. Depending on the length of linear dimethylsiloxane chain the copolymers are transparent or slightly opalescent sub-stances, soluble in ordinary organic solvents with $\eta_{sp}\approx 0,11\div 2,33$. Some physical-chemical properties and yields of copolymers listed in Table 8 and Table 9.

Table 8. Some physical-chemical characteristic of card-type copolymers with cyclocarbosiloxane fragments in the chain

№	Copolymer structure	n_{SiO}	η_{sp}	T_g, ^0C	$\bar{M}_{\omega}x10^3$	d_1, Å	Residual, mass, %	Yield, %
1	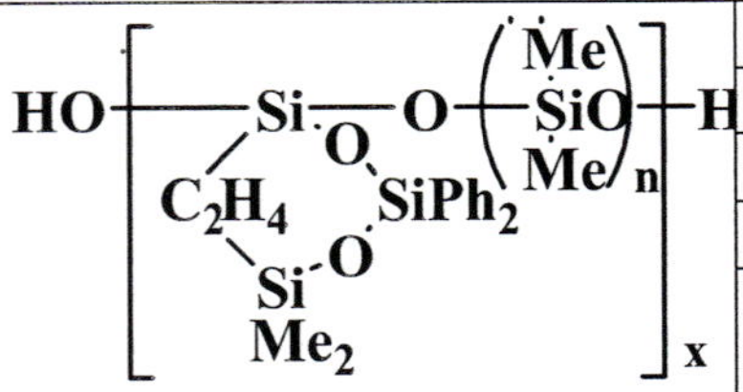	2	0,08	-51	78.4	9,50	46	74
2		4	0,12	-36	-	8,70	36	80
3		10	0,18	-85	-	-	-	89
4		21	0,20	-119	124	7,25	32	92
5		36	0,25	-123	-	7,25	-	95

Table 9. Some physical-chemical charecteristic of card-type copolymers with silaoxadihydrophenanthrene fragments in the chain

№	Copolymer structure	n_{SiO}	[η] dl/g	T_g, ^0C	5% mass losses	$\bar{M}_{\omega}x10^3$ (D)	d_1,Å	Yield, %
1	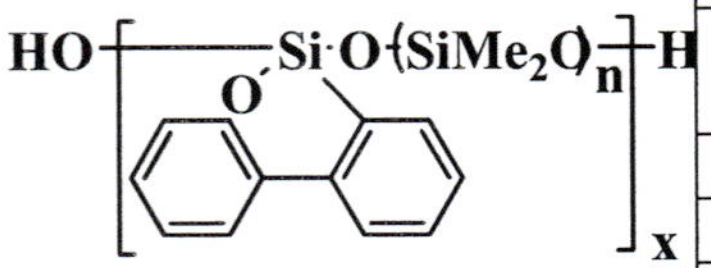	2	0,11	0	350	-	9,30	73
2		4	0,15	-32	300	150 (2.0)	8,23	86
3		8	0,28	-62	-	-	-	93
4		12	0,40	-85	-	-	-	94
5		19	1.82	-112	350	387 (2.6)	7,50	95
6		27	2.33	-123	300	648 (3.0)	7,50	94

During reprecipitation of copolymer 1 (Table 8 and 9) from the mother solution of toluene and methyl alcohol, after removal of solvents low-molecular weight product were obtained ~6%, which according to its composition and structure correspond to bicyclic structure XI and XII. For this product molecular masses, IR and NMR spectra were determined. On the basis of such results it was proved the cyclic structure of compounds with structure XI and XII. Such product of cyclic structure XII (n=2) was obtained directly by HFC of 1,1-dichloro-1-sila-2-oxadihydrophe-nanthrene with 1,3-dihydroxytetramethyl-disiloxane with 1:1 ratio of the initial compounds in the presence of aniline in 5% solution of anhydrous ether.

After distillation of the reaction products bicyclic compound with the structure XII has been obtained. The reaction proceeds according to the following scheme:

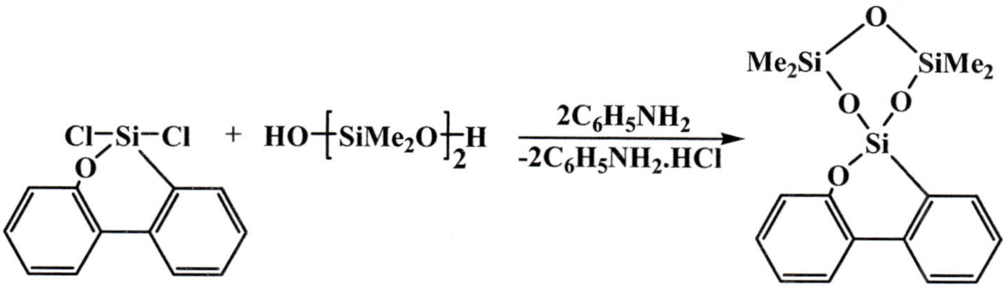

XII

(5.21)

The composition and structure of compound with the structure XII were determined by means of elementary analysis, by finding of molecular masses, by IR and ^{1}H NMR spectra data.

Copolymer 3 (Table 8) and 4 (Table 9) was fractionated and it was shown, that elementary analysis of obtained fractions slightly distinguished from each other and from initial copoly-mers. Which proved that during HFC secondary processes does not take place and the reaction mainly proceeds by HFC way.

GPC investigation of copolymers with the structure XIII has been carried out and it was shown, that MWD of copolymers is monomodalen. The molecular weights $\overline{M}_\omega$ of copolymers are in the range $1{,}5\text{x}10^5 \div 5{,}0\text{x}10^5$ and polydispersity of copolymers changes in the range $2{,}0 \div 3{,}0$. The values of molecular masses ($\overline{M}_\omega$) and polydispersity are presented in the Table 9.

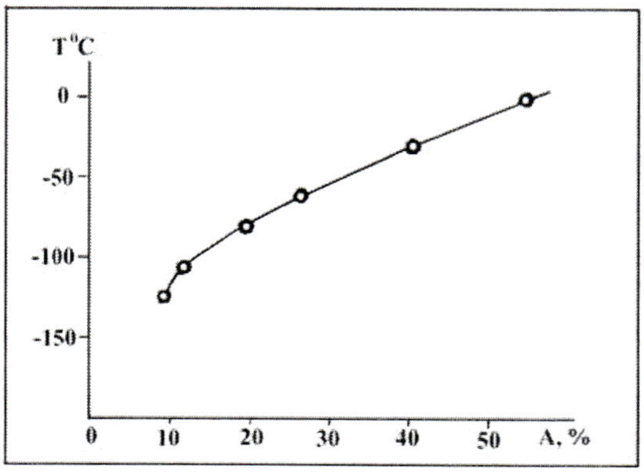

Figure 6. Dependence of Tg from mole % containing of silaoxadihydrophe-nanthrene (A) fragments.

For synthesized copolymers thermomechanical and DSC investigations has been carried out. As it's seen from the Tables 8 and 9 the maximum value of glass temperature T_g of copolymers reaches at a small lengths of the linear polydimethylsiloxane chain. On Figure 6 the dependence of T_g on the length of linear dimethylsiloxane chain is presented, from which this is evident that, with an increase of the length of dimethylsiloxane chain the T_g of copolymers XIII decreases until to T_g of the linear polydimethylsiloxane. The influence of cyclocarbosiloxane and 1-sila-2-oxadi-hydrophenanthrene fragments on the length of dimethylsiloxane chain is sensible until n≈19-21, in this case T_g of copolymers 4 (Table 8) and 5 (Table 9) is equel to –112÷-119 ^{0}C.

The copolymers are one phase amorphous system and maximum values of interchain distances d_1 is observed at a small length of dimethylsiloxane chain and for copolymer 1 d_1≈9,30-9.50 Å (Table 8 and 9), this value of a d_1 approximetely equals to d_1 characteristic for cyclolinear copolymers with organocyclopenta(hexa)siloxane and cyclocarbosiloxane fragments in the main dimethylsiloxane chain[42,43]. As it's seen from the Table 8 and 9 data an increase of the length of dimethylsiloxane chain n the value of d_1 decreases and for copolymer 5, d_1≈7,25 Å, this value of interchain distances equal to d_1 characteristic for the linear PDMS.

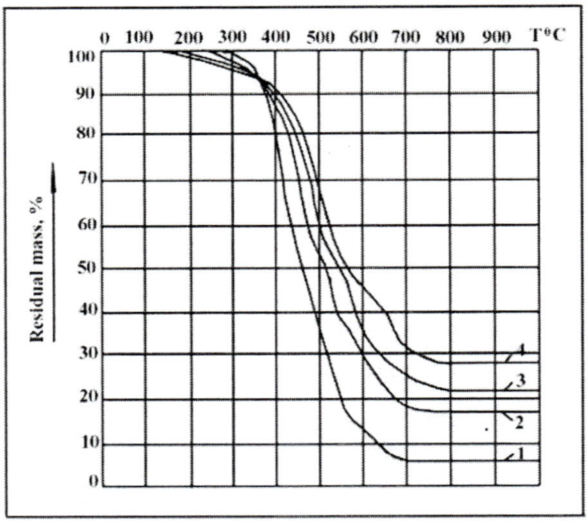

Figure 7. The thermogravimetric curves for copolymers with silaoxadihydrophenanthrene fragments in dimethylsiloxane chain. Where curve 1 corresponds to copolymer 1, curve 2-2, 3-5 and 4-6, (Table 9)

The thermogravimetric investigation for copolymers with the structure X and XIII has been carried out. It was shown that card-type copolymers with cyclocarbosiloxane and silaoxadihyd-rophenanthrene fragments in the chain are characterized approximately by the same thermal-oxi-dative stability. On Figure 7 the thermogravimetric curves of copolymers with the structure XIII are presented. The copolymers with a small length of dimethylsiloxane chain are characterized by higher thermal-oxidative stability (curves 1 and 2), with an increase of the lenght of dimethyl-siloxane chain thermal-oxidative stability of copolymers decreases. 5% mass losses for copolymers are observed in the temperature range 300÷350^{0}C, the main destruction process for both type co-polymers proceeds in termperature 450÷700^{0}C range and above 700^{0}C the curves of mass losses do not changes. So, with the insertion of card-type heterocyclic fragments in dimethylsiloxane chain agrees on literary dates [45,46] break the spiral structure of copolymers and changes the physical-chemical properties of copolymers. The main process of thermal-oxidative destruction starts about 80-100^{0}C higher compared with the unblocked linear PDMS [47].

5. Synthesis and Investigation of Properties of Card-Type Siliconorganic Oligomers with Silaoxadihydrophenentrene Fragment in the Chain

For the purpose of synthesis of silaoxadihydrophenanthrene-diphenylsiloxane containing di-hydroxy end group oligomers the reactions of heterofunctional condensation of 1,1-dichloro-2-oxa-dihydrophenanthrene with dihydroxydiphenylsilane at 1:0,9 and 1:0,95 ratio of initial compounds in the presence of dry pyridine in 50-60% solution of anhydrous toluene have been carried out. The reaction proceeds according to the following scheme [48,49]:

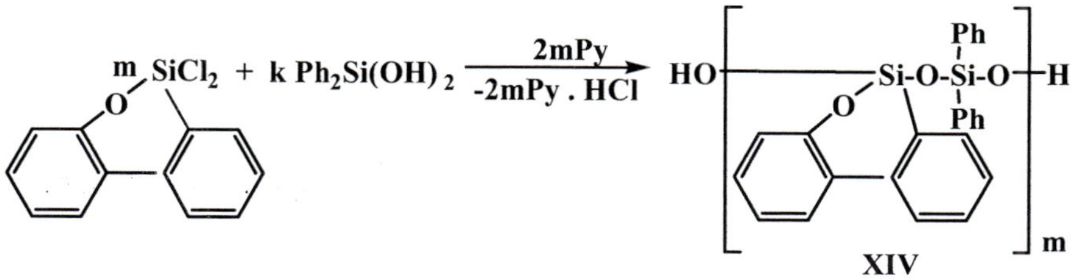

XIV

(5.22)

Where: m≈10 (I), 14(II).

After reprecipitation of reaction products from toluene solution by methanol white solid type α,ω–dihydroxysilaoxadihydrophenanthrene-diphenylsiloxane oligomers soluble in ordinary organic solvents have been obtained.

The composition and structure of synthesized oligomers were determined by means of ele-mentary and functional analysis, by finding of molecular masses, by IR spectra data. Some physi-cal-chemical properties of synthesized oligomers are presented in Table 10.

Table 10. Some physical chemical properties of α,ω–dihydroxysilaoxadihydro-phenanthrene-diphenylsiloxane oligomers (structure XIV)

№	Structure	η^*_{sp}	d_1,Å	T_{melt} ^{0}C	OH,%	$\bar{M}$**	Yield,%
1	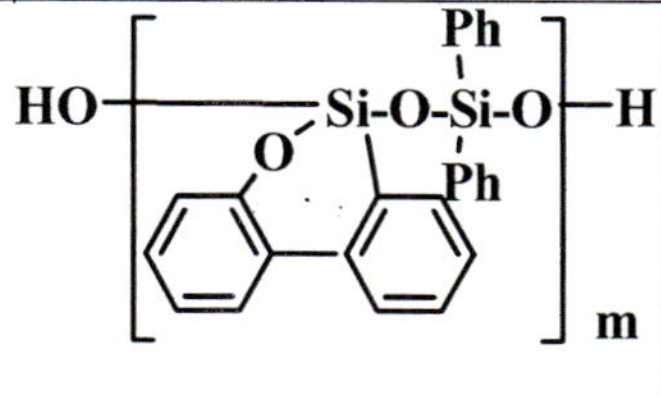	0,04	11,05	67÷72	0,83 / 0,81	4111 / 4200	92
2		0,05	11,05	74÷77	0,59 / 0,56	5758 / 6070	93

*In 1% toluene solution at 25^{0}C. **In numerator calculated values, in denominator experimental values.

For the purpose of synthesis of card-type oligomers with silaoxadihydrophenantrene fragments in silarylene chain the reaction of dehydrocondensation of 1,1-dihydroxy-1-sila-2-oxadihyd-rophenantrene (compound 2, Table 5) with 1,4-bis(dimethylsilyl)benzene has been investigated.

The reaction was carried out at 1:1 ratio of initial compounds in the presence of anhydrous caustic potassium as a catalyst (0,1% weight), in the temperature range $40\div60^0$C in anhydrous dimethylformamid (C≈0,3952 mole/l). It was established that in this conditions no changes of initial compounds and the polymerization of silaoxadihydrophenanthrene fragments does not takes place [50].

Dehydrocondensation reaction of 1,1-dihydroxy-1-sila-2-oxadihydrophenantrene with 1,4-bis(dimethylsilyl)benzene proceeds according to the following scheme [50,51]:

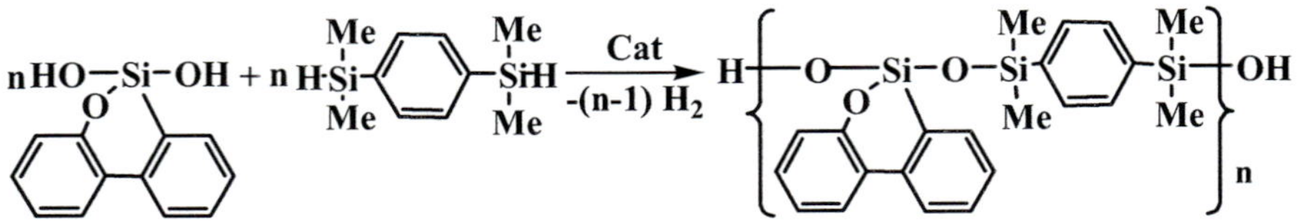

(5.23)

XV

where: at 40^0C (1^1), 50^0C(1^2), 60^0C(1^3)

The synthesized oligomers has been reprecipitated from toluene solution by methanol and they are represented as solid yellow colour products well soluble in ordinary organic solvents with η≈0,09-0,11. It was established that with an increase of the temperature the reaction rate rises. Figure 8 shows hydrogen conversion in dehydrocondensation reaction of dihydroxysilaoxadihyd-rophenanthrene with 1,4-bis(dimethylsilyl)benzene.

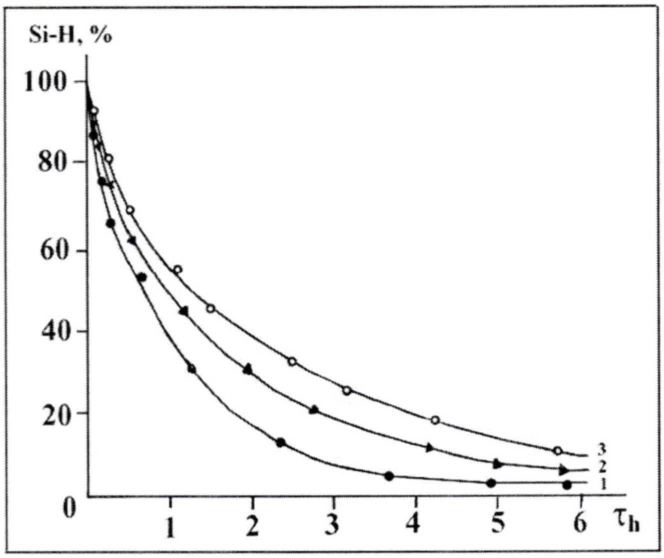

Figure 8. Hydrogen conversion in dehydrocondensation reaction of dihydroxysilaoxadihydrophenanthrene with 1,4-bis(dimethylsilyl)benzene. Where curve1 corresponds to a reaction temperature at 60^0C, curve 2 – 50^0C and curve 3- 60^0C.

It was established that the dehydrocondensation reaction is the reaction of second order and from the dependence of reverse concentration on the time, the reaction rate constants at various temperature were calculated: $k_{40^0C} \approx 3,3725 x10^{-2}$; $k_{50^0C} \approx 4,6625 \ x10^{-2}$; $k_{60^0C} \approx 6,3995 x10^{-2}$. The acti-vation energy of the dehydrocondensation reaction is $\Sigma \approx 29,44$ KJ/mole. By finding of molecular masses of oligomer 1^3 it was shown that the dehydrocondensation reaction proceeds more deeply (for copolymer 1^3 $m \approx 17$) than heterofunctional condensation of 1,1-dichloro-1-sila-2-oxadihydro-phenanthrene with 1,4-bis(hydroxydimethylsilyl)benzene (see reaction 23). Some physical-chemi-cal characteristics of oligomers with the structure XV (reaction 22) are presented in Table 11.

Table 11. Some physical-chemical characteristics of oligomers with the structure XV (reaction 22)

№	Oligomer structure	Reac-tion T^0,C	$\eta_{sp.}$	d_1, Å	Conver-sation H_2, %	m	T_{soft}, ^0C	Yield, %
1^1	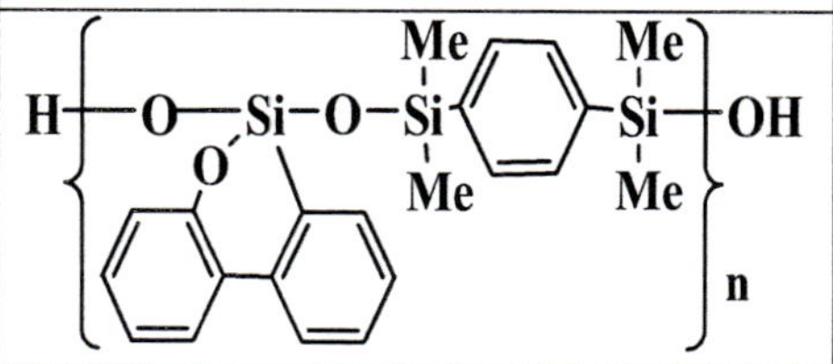	40	0.09	10.74	93	-	64-68	93
1^2		50	0.10	-	95	-	65-70	95
1^3		60	0.11	10.74	98	17	67-74	98

For the purpose of synthesis of silaoxadihydrophenanthrene-silarylen oligomers, containing hydroxyl groups at the end, the reaction of HFC of 1,1-dichloro-1-sila-2-oxadihydrophenanthrene (compound 1, Table 5) with 1,4-bis(hydroxydimethylsilyl)benzene at various ratio of initial com-ponents (1;0,95 and 1:1) in 50-60% solution of anhydrous toluene in the presence of pyridine has been carried out. The reaction of HFC proceeds according to the following scheme [52,53]:

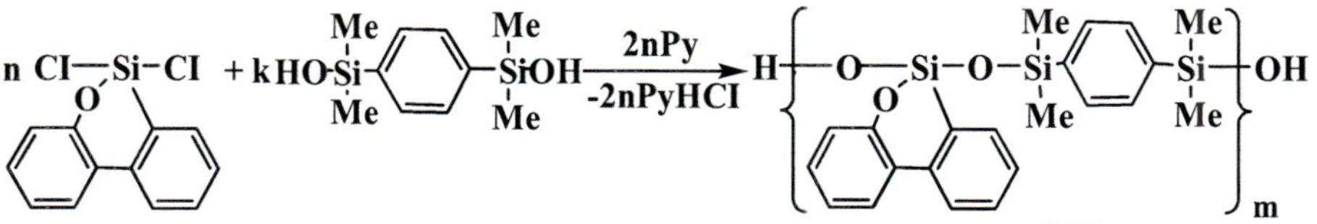

XV (5.24)

where: $m \approx 6 \div 12$.

The synthesized oligomers has been reprecipitated from toluene solution by methanol and they are represented as solid white-yellow colour products well soluble in ordinary organic solvents with $\eta \approx 0,06$-$0,09$. Some physical-chemical properties of synthesized silarylene-silaoxadihydro-phenanthrene oligomers are presented in Table 12.

Thermogravimetric investigation for synthesized silarylen-silaoxadihydrophenanthrene oligomers has been carried out. For silarylen-silaoxadihydrophenanthrene oligomer 10% mass losses one can observe in 340-360^{0}C temperature range. The main destruction process proceeds over the range 450-680^{0}C and above 700^{0}C the curve of mass losses does not change. By roentgenographic investigation it was shown that oligomers with the structure XV are one phase amorphous systems with the value of interchain distances $d_1 \approx 10{,}74$ Å. The obtained oligomers are interesting products for synthesis of $(AB)_nC_m$ type block-copolymers.

Table 12. Some physical-chemical properties of synthesized silarylene-silaoxadihydrophenanthrene oligomers (reaction 24)

№	Oligomer	$\eta_{sp.}$	d_1, Å	OH,%	m	T_{soft}, ^{0}C	Yield, %
1	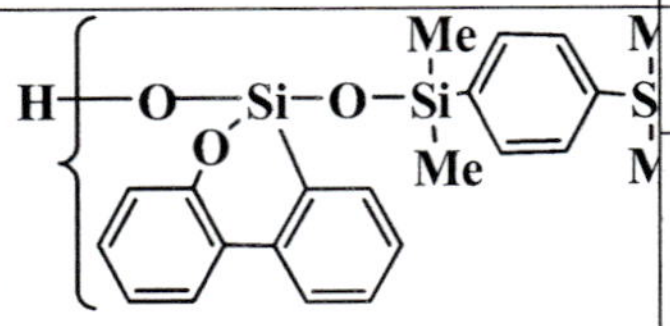	0.06	10.74	1.34 / 1.32	6	58-65	89
2		0.09	10.74	0.67 / 0.66	12	68-71	92

*In numerator there are calculated values, in denominator – experimental values.

Thus, a conclusion can be made that if introduced to the macromolecular chain, cardocyclic fragments behaves similar to cyclic fragments in the backbone or in the side framing. By analogy with cyclic fragments in dimethylsiloxane backbone, they inhibit the chain transfer reaction during degradation and the main degradation process is initiated 60 - 80°C higher, than for linear poly(dimethylsiloxanes).

REFERENCES

[1] Wu T.C. and Hirt C.A., *J. Organomet. Chem.*, 1968, vol. 11(1), p. 17.

[2] Andrianov K.A. and Zachernyuk A.B., *Zh. Vses. Khim. Obshch. im. D.I. Mendeleeva*, 1974, vol. 19(1), p. 107. (Rus)

[3] Zachernyuk A.B., Burlova E.A., and Zhdanov A.A., *Zh. Obshch. Khim.*, 1985, vol. 55(6), p. 1368. (Rus)

[4] Meladze S.M., Mukbaniani O.V., Makarova N.N., and Khananashvili L.M., *Zh. Obshch. Khim.*, 1980, vol. 50(11), p. 2493. (Rus)

[5] Andrianov K.A., Zachernyuk A.B., and Danielov S.A., *KhGS*, 1972, No. 7, p. 893. (Rus)

[6] Andrianov K.A. and Zachernyuk A.B., *KhGS*, 1972, No. 9, p. 1183. (Rus)

[7] Zachernyuk A.B., *Candidate's Dissertation on Chemistry*, 1974, INEOS AS USSR, Moscow. (Rus)

[8] Andrianov K.A., Zachernyuk A.B., and Lavrukhin B.D., *Doklady AN SSSR*, 1972, vol. 207(3), p. 593. (Rus)

[9] Andrianov K.A., Zachernyuk A.B., Lavrukhin B.D., and Vasilenko N.G., *Doklady AN SSSR*, 1972, vol. 204(5), p. 1112. (Rus)

[10] Hicton H.J., Holt A. J. Chem., Soc., 1966, p.149.

[11] Andrianov K.A., Zachernyuk A.B., and Zhdanova E.A., *Doklady AN SSSR*, 1974, vol. 214(2), p. 325. (Rus)

[12] Andrianov K.A., Zachernyuk A.B., and Solomatin G.V., *Doklady AN SSSR*, 1977, vol. 236(3), p. 601. (Rus)
[13] Zachernyuk A.B. and Solomatin G.V., *Thes. Rep. V All-Union Conference on Chemistry and Application of Organosilicon Compounds*, 1980, January 29 – February 1, Tbilisi, part 1, p. 281. (Rus)
[14] Vdovin V.Sh. and Petrov A.D., *Zh. Obshch. Khim.*, 1960, vol. 30, p. 838. (Rus)
[15] Petrashko A.I., Andrianov K.A., Korneeva G.P., Kuntsova Z.M., Moiseenko A.P., and Asnovich A.Z., *Vysokomol. Soedin.*, 1967, vol. 9, p. 2036. (Rus)
[16] Inaridze I.A., *Candidate's Dissertation on Chemistry*, I. Javakhishvili Tbilisi State University, 1993, Tbilisi, Georgia.
[17] Chernyishev E.A., Krasnova T.L., and Shchepinov S.A., *Zh. Obshch. Khim.*, 1970, vol. 40, p. 1735. (Rus)
[18] Chernyishev E.A., Krasnova T.L., and Shchepinov S.A., *Zh. Obshch. Khim.*, 1970, vol. 40, p. 1058. (Rus)
[19] Dewar M.J.S., *J. Amer. Chem. Soc.*, 1960, p. 1344.
[20] Dewar M.J.S. and Cubba P., *J. Amer. Chem. Soc.*, 1959, p. 347.
[21] Samsonia A.Sh., Mukbaniani O.V., and Karchkhadze M.G., *Georgian Eng. News*, 2000, No. 2, p. 144.
[22] *A.C. No. 602,510*, 1976, Andrianov K.A., Zachernyuk A.B., and Burlova E.A.; *Bull. Izobr.*, 1978, No. 14, p. 85. (Rus)
[23] Stabilization of Polymers and Stabilizing Processes. Advances in Chemistry, Series, *Amer. Chem. Soc.*, Washington, 1968.
[24] Andrianov K.A., *High-molecular Compounds for Electrical Insulation*, Gosenergoizdat, 1961. (Rus)
[25] Andrianov K.A. and Zachernyuk A.B., *Vysokomol. Soedin.*, 1974, vol. 16B(6), p. 307. (Rus)
[26] Andrianov K.A., Burlova E.A., and Zachernyuk A.B., *Vysokomol. Soedin.*, 1977, vol. 19B(6), p. 211. (Rus)
[27] Andrianov K.A. and Zachernyuk A.B., *Vysokomol. Soedin.*, 1974, vol. 16A(6), p. 1435. (Rus)
[28] Bunina L.O., Burlova E.A., and Zachernyuk A.B., *Thes. Rep. VII All-Union Conference on Chemistry, Production Technology and Application of Organosilicon Compounds*, 1990, Tbilisi, part 2, p. 405. (Rus)
[29] *Patent No. 3,366,593* (USA), 1969, Breed L.W. and Elliott L.L.
[30] Andrianov K.A., Lekishvili N.G., Nogaideli A.I., Kotrelev G.V., and Tkeshelashvili R.Sh., *Vysokomol. Soedin.*, 1974, vol. 16B, p. 496. (Rus)
[31] Andrianov K.A., Nogaideli A.I., Kotrelev G.V., Lekishvili N.G., Kamarnitski B.A., and Kryilov A.P., *Vysokomol. Soedin.*, 1975, vol. 17B, p. 315. (Rus)
[32] Andrianov K.A., Kotrelev G.V., Nogaideli A.I., Zhuravleva I.V., Lekishvili N.G., Tolchinski Yu.P., and Pushich V.I., *Vysokomol. Soedin.*, 1977, vol. 19A, p. 451. (Rus)
[33] Kotrelev G.V. and Zhdanova E.A., *Vysokomol. Soedin.*, 1981, vol. 23B(4), p. 289. (Rus)
[34] Inaridze I.A., Mukbaniani O.V., and Khananashvili L.M., *Bulletin of the Georgian Academy of Sciences*, 1994, vol. 149(1), p. 84.
[35] Mukbabiani O.V., *Doctor's Dissertation on Chemistry*, I. Javakhishvili Tbilisi State University, 1993, Tbilisi, Georgia. (Rus)

[36] Khananashvili L.M., Mukbaniani O.V., Samsonia A.Sh., and Karchkhadze M.K., *XVI Mendeleev Meeting on General and Applied Chemistry*, Saint-Petersburg, May 25 – 26, 1998, p. 320. (Rus)

[37] Mukbaniani O.V., Scherf U., Samsonia A.Sh., Karchkhadze M.G., and Khananashvili L.M., *Intern. J. Polym. Mater.*, 2001, vol. 48(4), p. 381.

[38] Andrianov K.A., Nogaideli A.I., Makarova N.N., and Mukbaniani O.V., *Izv. AN SSSR*, 1977, No. 6, p. 1388. (Rus)

[39] Yuzhelevski Y.A., Kurlova T.V., Kagan E.A., and Suvorova M.V., *Zh. Obshch. Khim.*, 1972, No. 9, p. 2006.

[40] Andrianov K.A., Nogaideli A.I., Slonimski G.L., Levin V.Yu., Kvachev Yu.P., Makarova N.N., and Mukbaniani O.V., *Vysokomol. Soedin.*, 1976, vol. 13B(5), p. 359. (Rus)

[41] Meladze S.M., Khananashvili L.M., and Mukbaniani O.V., *Vysokomol. Soedin.*, 1984, vol. 26B(5), p. 250. (Rus)

[42] Koyava N.A., Mukbaniani O.V., and Khananashvili L.M., *Vysokomol. Soedin.*, 1985, vol. 27A(11), p. 2261. (Rus)

[43] Mukbaniani O.V., Achelashvili V.A., Levin V.Yu., Meladze S.M., Inaridze I.A., and Khananashvili L.M., *Izv. AN GSSR*, 1990, vol. 16(1), p. 20. (Rus)

[44] Mukbaniani O.V., Achelashvili V.A., Khananashvili L.M., and Sturua G.I., *Izv. AN GSSR*, Coll., 1988, p. 53. (Rus)

[45] Thomas T.H. and Kendrik T.C., *J. Polym. Sci.*, 1969, vol. 2A, p. 537.

[46] Alexander G. and Garzo G., *Chromatography*, 1974, vol. 7, p. 225.

[47] Andrianov K.A., Papkov V.S., Slonimski G.L., Zhdanov A.A., and Iakushina S.E., *Vysokomol. Soed.*, 1969, vol. 11A, p. 2030. (Rus)

[48] Mukbaniani O.V., Samsonia A.Sh, Karchkhadze M.G., and Khananashvili L.M., J. Appl. Polym. Sci., 2002, vol. 84, p. 9.

[49] Samsonia A.Sh., Karchkhadze M.G., Mukbaniani O.V., and Khananashvili L.M. *Georg. Eng. News*, 2000, No. 2, p. 144.

[50] Samsonia A.Sh., Mukbaniani O.V., and Karchkhadze M.G., *Georg. Eng. News*, 2000, No. 2, p. 142.

[51] Samsonia A.Sh., *Candidate's Dissertation on Chemistry*, I. Javakhishvili Tbilisi State University, 2002, Tbilisi, Georgia. (Rus)

[52] Mukbaniani O.V., Samsonia A.Sh., and Karchkhadze M.G., *Condensation Polymers: Synthesis, Structure, Properties*, (To the 90th anniversary of Academician V.V. Korshak), January 12 – 14, 1999, Moscow, pp. 1 – 15. (Rus)

[53] Samsonia A.Sh., Karchkhadze M.G., and Mukbaniani O.V., *Georg. Eng. News*, 2000, No. 2, p. 142. (Rus)

In: Frontiers in Physical Organic Chemistry
Editor: G.E. Zaikov, pp. 85-102
ISBN 1-60021-128-3

Chapter 6

THE CONCEPT OF STRUCTURAL DEFECTS WITHIN THE CLUSTER MODEL

G. V. Kozlov and G. E. Zaikov[*]
Kabardino-Balkarian State University, 173, Chernychevskogo st., Kabardino-Balkaria, Nal'chik, 360004, Russia
[*]N.M.Emanuel Institute of Biochemical Physics, Russian Academy of Sciences, 4, Kosygin st., Moscow, 119991, Russia

As mentioned in [1], the cluster model of the amorphous polymer structure allows introduction of principally new interpretation of the structure defect (with the vengeance of this term) for the mentioned state [1, 2]. The essence of this model will be discussed in detail in this paper. It is common knowledge [3] that structure of natural solids contains a significant amount of defects. This concept forms the basis for the dislocation theory, widely applied to description of crystalline solids' behavior. Success in this branch has predetermined attempts of many authors (for example, of works [4 – 11]) to apply this concept to amorphous polymers. Hence, the notions used for crystalline lattices are often transposed to the structure of amorphous polymers. Usually, the basis for this transposition is formal resemblance of stress – deformation (σ - ε) curves for crystalline and amorphous solids. It has been suggested [2] that the defect theory of polymer amorphous state must be based not on the formal resemblance of behavior of the mentioned classes of solids, but on the principal difference in their structures.

In relation to the structure of amorphous polymers, for a long time the most ambiguous point [12 - 14] was the presence or the absence of the local (short range) order in which connection points of view of different authors on this problem were significantly different. The presence of the local order can significantly affect determination of the structure defect in amorphous polymers, if in the general case the order-disorder transition or vice versa is taken for the defect. For example, any distortion (discontinuity) of the long range order in crystalline solids represents a defect (dislocation, vacancy, etc.), and a monocrystal with the proper long range order is the ideal defect-free structure. It is known [15] that sufficiently bulky sample of 100% crystalline polymer cannot be obtained, and all characteristics of such hypothetical polymers are determined by the extrapolation method. That is why the Flory "felt" model [16, 17] can be suggested as the ideal defect-free structure for amorphous polymers. This model suggests that amorphous polymers consist of interpenetrating macromolecular coils interpreting the full disorder (chaos). On this basis, for the defect in amorphous polymers a distortion (discontinuity) of full disorder must be accepted, i.e. formation of the local (or long range) order [2]. It shall be noted also that formal resemblance of the σ - ε curves for crystalline solids and amorphous polymers appears incomplete and behavior of these classes of materials displays principal differences, which will be discussed in detail below.

Turning back to the suggested concept of amorphous polymer structural defect, let us note that a segment participating in the cluster can be considered the linear defect – the analogue of dislocation in crystalline solids. Since in the cluster model the length of such segment is accepted equal the length of statistical segment, l_{st}, and their amount in the specific volume equals frequency of the cluster entanglement network, ν_{cl}, the density of linear defects, ρ_d, in the specific volume of the polymer is presented by the following relation [2]:

$$\rho_d = L_{cl} = \nu_{cl} l_{st} \quad . (1)$$

Suggested interpretation allows application of well developed mathematical apparatus of the dislocation theory to description of amorphous polymer properties. This point will be thoroughly discussed below.

With respect to suggested concept of structural defect in amorphous polymers, of special attention are results of wide-angle X-ray analysis which indicate an antibate change of the relative part of clusters (the local order degree), φ_{cl}, and integral intensity (the area occupied by the amorphous halo) I^{in} of total amorphous halo as the function of cross-linking frequency, ν_c, for epoxy polymers [18]. If the consideration is based on the analogy with crystalline materials, this result seems unexpected, because φ_{cl} increase should induce consequent I^{in} growth [19]. Simultaneously, this observation is completely explained by the structure defect concept, suggested in this consideration. In this case, the disorder exponent in the system (i.e. the degree of ideal defect-free structure approaching) can be presented by the relative part of the fluctuation free volume in the glassy state, f_f [20, 21].

Figure 1 shows the dependence $I^{in}(f_g)$ which is linear and passes through the origin of coordinates. To put it differently, for polymer amorphous state I^{in} is the value characterizing the disorder degree that explains the above-mentioned antibate type of dependencies $\varphi_{cl}(\nu_c)$ and $I^{in}(\nu_c)$ and proves the suggested concept of defect, as well. Generally, an analogous interpretation was suggested by Ginnier [22]. He has assumed that for metals I^{in} can be decreased due to formation of pre-crystallite structures in multicomponent melts.

Computerized expansion of wide-angle X-ray diffraction patterns into two components of packless matrix and clusters was performed [18]. Figure 2 shows an example of such expansion for PC. This procedure allowed separate comparison of characteristics of the amorphous halo appropriate components and parameters of the cluster structure components (Figure 3).

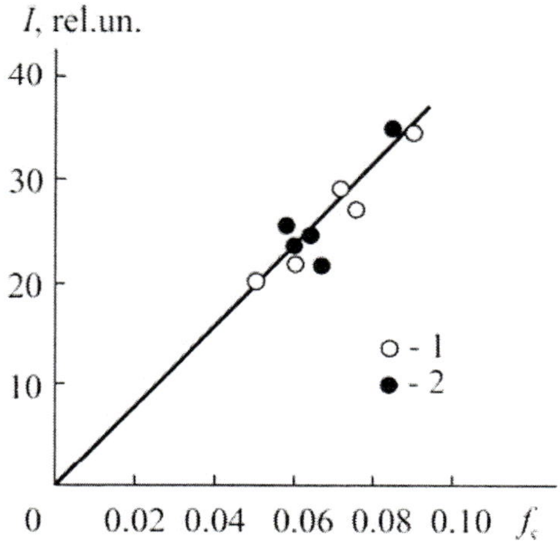

Figure 1. The relation between integral intensity, I^{in}, of wide-angle X-ray patterns and relative fluctuation free volume, f_g, for epoxy polymers of amine (1) and anhydride (2) cross-linking type [2]

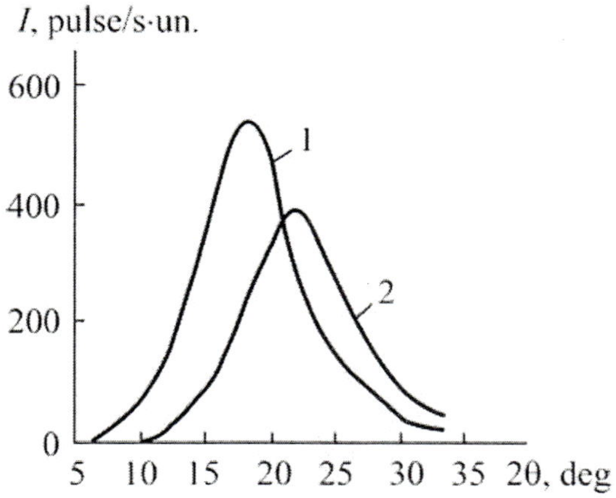

Figure 2. Expansion of experimental wide-angle diffraction pattern for PC into components: halo for packless matrix (1) and clusters (2) [18]

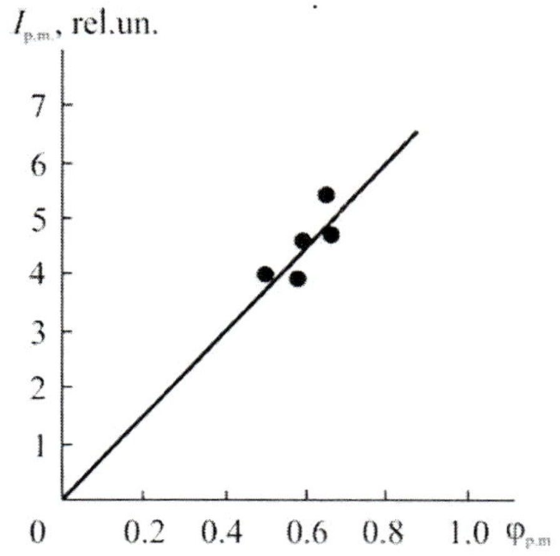

Figure 3. Dependence of integral intensity of packless matrix dispersion component, $I_{p.m.}$, on its relative part, $\varphi_{p.m.}$, for epoxy polymers of the amine cross-linking type [18]

Comparison of the results shown in Figures 1 and 3 with appropriate data on amorphous polymers from the literature indicates that the above-states regularities do not contradict with phenomena, observed by other researchers. For example, the study of annealing temperature (T_{an}) effect on cellulose nitrate structure [23] has detected systematic decrease of amorphous halo intensity and increase of the Bragg interval (d) with temperature T_{an}. It is commonly accepted that annealing of amorphous polymers at temperature below T_g improves molecular packing [24, 25]. As a consequence, this process must be intensified with T_{an} increase. Though different explanation of the effect observed has been suggested [23], experimental data from this work fit the above-suggested interpretation. Decrease of I^{in} with cluster formation was also observed in the work [26].

Generally, it is expected that with temperature increase a solid becomes more disordered. The more so, if similar to the cluster model the order is of thermofluctuational origin, this tendency becomes more obvious. However, studies of inorganic glasses (for example, As_2Se_2 and GeS_2) using X-ray diffraction analysis and small-angle neutron scattering methods indicate that the peak intensity related to the short range order is increased in the amorphous state with temperature [19]. This allowed an assumption that the layered structure embodying the short range order in these glasses becomes more regular with temperature. It should be noted that besides the above-mentioned analogies with the behavior of crystalline solids, no other proofs were presented [19]. The interpretation suggested explains this experimental observation more realistically. Moreover, on the example of different glasses [19], it is indicated that the percentage increase of the peak intensity mentioned is as higher as closer temperature approaches T_g of glass. This observation also correlates with the cluster model. Finally, the increase of amorphous halo intensity with temperature both above and below T_g was also observed for PS [27]. Obviously, there are no grounds to suggest intensification of the local order with temperature for this amorphous polymer [28]. That is why it shall be admitted that suggested interpretation of the structural defect gives just a general explanation of experimental results [19, 23, 26 – 28].

An interesting experimental result was obtained for tensile stressed epoxy polymers [29]. As deformation of the epoxy polymer exceeds its strain yield, intensity of the amorphous halo was increased approximately by 17%. This result may also be easily explained in the framework of the interpretation suggested, because as shown below the cluster model suggests that yielding is associated with decomposition of unstable clusters and, consequently, with decrease of ν_{cl} and the local order degree [30, 31]. Concentration of clusters of this type is approximately corresponded to the increase of amorphous halo intensity observed for the epoxy polymer [29, 31].

Intensity of the halo second component (Figure 2, curve 2), associated with clusters (I_{cl}), displays a linear correlation I_{cl}^{-1} (φ_{cl}) (Figure 4). The latter is coordinated with the above-stated ideas on the intensity decrease with ordering increase in amorphous polymers.

Further on, the rightfulness of application of the structural defect concept suggested to polymers wielding will be discussed. As a rule, previously suggested concepts of defects in polymers were primarily used for description of this process or even exclusively for this purpose [4 – 11]. Theoretical shear strength of crystals was primarily calculated by Frenkel. He has based on a simple model of two series of atoms, shifted in relation to one another by the shear stress (Figure 5a) [3]. According to this model, critical shear stress τ_0 is expressed as follows [3]:

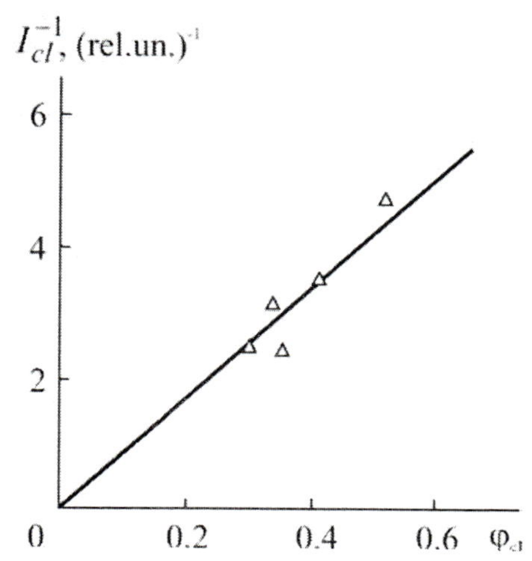

Figure 4. Dependence of reverse scattering curve intensity from clusters $\left(I_{cl}^{-1}\right)$ on their relative fracture (φ_{cl}) for epoxy polymers of the amine cross-linking type [18]

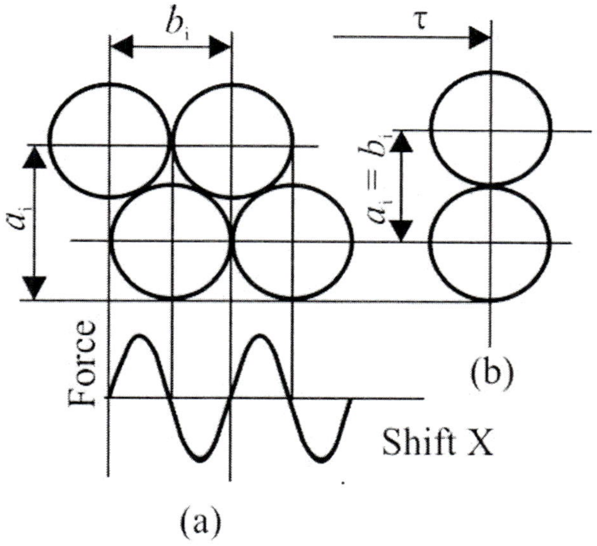

Figure 5. Schematic representation of deformation of two sequences of atoms in accordance with the Frenkel model. Positions before (a) and after (b) deformation [3]

$$\tau_0 = \frac{G}{2\pi} \tag{2}$$

where G is the shear modulus.

Slightly changed, this model was used in the case of polymer yielding [6], wherefrom the following relation was obtained:

$$\tau_{0,y} = \frac{G}{\pi\sqrt{3}} \tag{3}$$

where $\tau_{0,y}$ is theoretical value of the shear stress for yielding.

Special attention should be paid to the fact that characterizes principally different behavior of crystalline metals compared with polymers. As is known [3, 32], $\tau_{0,y}/\tau_y$ relation (where τ_y is experimentally determined shear stress for yielding) is much higher for metals than for polymers. For five metals possessing the face-centered cubic or hexagonal lattices the following relations were obtained: $\tau_{0,y}/\tau_y = 3{,}740 - 22{,}720$ [3], whereas for five polymer this relation equaled 2.9 - 6.3 [6]. As a matter of fact, sufficient closeness of $\tau_{0,y}/\tau_y$ relation values to unit may already be the proof for possibility of realizing the Frenkel mechanism in polymers (in contrast with metals), but it is shown below that for polymers, a small modification of the law of shear stress regular change (τ) used commonly gives $\tau_{0,y}/\tau_y$ values very close to unit [33].

As is shown [34], dislocation analogies are also true for amorphous metals. Essentially, distortion of the composition of atoms (which induces occurrence of elastic stress fields) is considered [34] as a linear defect (dislocation) being practically immovable. Clearly such approach correlates completely with the structural defect concept, suggested in this paper. In the framework of this concept, Figure 5a may be considered as a cross-section of a cluster (crystallite). As a consequence, the shift of segments in this cluster, according to the Frenkel mechanism, can be taken for the one limiting yielding in polymers. This is proved by the experimental data [35] which show that glassy polymer yielding is realized directly in closely packed zones. Other data indicate [36] that these closely packed zones are clusters. To put it differently, one can state that yielding is associated with clusters (crystallites) stability loss in the shear stresses field [37].

Figure 6 shows asymmetrical periodical function [38] that displays dependence of shear stress (τ) on shear deformation (γ_{sh}). As shown before [32], asymmetry of this function and corresponded decrease of the energy barrier height determined by segments of macromolecules at the elementary yielding action is stipulated by formation of fluctuation free volume cavities during deformation (which is the specific feature of polymers [39]). Data in Figure 6 indicate that in the initial part of periodical curve from zero to the maximum dependence of τ on displacement x can be simulated by a sine function with the period shorter than in Figure 5. In this case, function $\tau(x)$ can be reduced to the following form:

$$\tau = k \sin\left(\frac{6\pi x}{b_i}\right) \tag{4}$$

This expression is fully corresponded to the Frenkel conclusion, except for randomly chosen numerical coefficient in brackets (6 instead of 2).

Future calculation of τ_0 by the method, described in ref. [3], and its comparison with experimental values of τ_y indicated their tight correspondence for nine amorphous and amorphous-crystalline polymers (Figure 7). This proves the possibility of realization of the above-suggested yielding mechanism at the segment level.

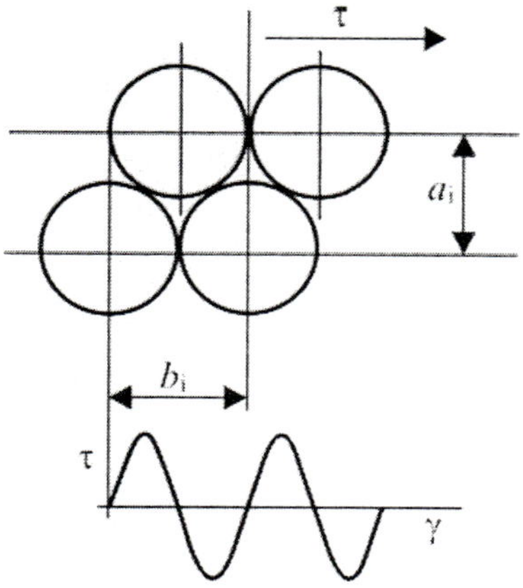

Figure 6. Schematic image of shear deformation and corresponded stress – deformation (τ - γ) function [38]

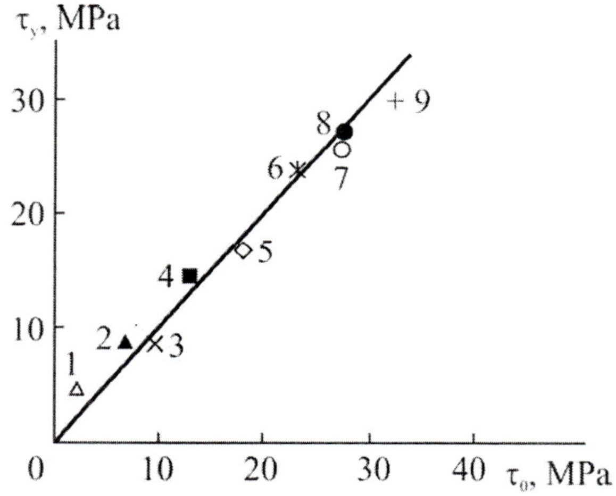

Figure 7. Correlation between theoretical (τ_0) and experimental (τ_y) shear stresses for yielding of polytetrafluoroethylene (1), HDPE (2), PP (3), polyamide-6 (4), PVC (5), PHE (6), PC (7), PSP (8) and PAr (9) [33]

Inconsistency of $\tau_{0,y}$ and τ_y values in metals induced a search for another mechanism of yielding realization. At present, it is commonly accepted that this mechanism is motion of dislocations by sliding planes of the crystal [3]. This implies that interatomic interaction forces, directed transverse to the crystal sliding plane, can be overcome in the case of presence of several local displacements, determined by periodical field of stresses in the lattice. This is strictly different from microscopic shift, during which all bonds are broken simultaneously (the Frenkel model). It seems obvious that with the help of dislocations total shear deformation will occur at applying much lower external stress than for the process including simultaneous break of all atomic bonds by the sliding plane [3].

Payerls and Nabarro [3] were the first who calculated the shear stress necessary for the dislocation motion, τ_{dm}. They used a sinusoidal approximation and deduced the expression for τ_{dm} as follows:

$$\tau_{dm} = \frac{2G}{1-\nu} e^{-2\pi a_i / b_i (1-\nu)} \tag{5}$$

where ν is the Poisson coefficient and parameters a and b are of the same meaning as in Figure 6.

By substituting reasonable value of ν, for example, 0.35 [40], and assuming $a_i = b_i$, the following value for τ_{dm} is obtained: $\tau_{dm} = 2\times10^{-4}$ G. Though for metals this value is higher than observed τ_y, it is much closer to them than the stress calculated using simple shear model (the Frenkel model, Figure 5).

However, for polymers the situation is opposite: analogous calculation indicates that their τ_{dm} does not exceed 0.2 MPa, which is by two orders of magnitude, approximately, lower than observed τ_y values.

The next point of discussion is the free path length of dislocations, λ_d. As known for metals [3], in which the basic role in elastic deformation is belonged to the mobile dislocations, λ_d is assessed as ~10^4 Å. For polymers, this parameter is estimated as follows [41]:

$$\lambda_d = \frac{\varepsilon_y}{b\rho_d} \tag{6}$$

where ε_y is the yielding deformation; b is the Burgers vector; ρ_d is the density of linear defects, determined from the equation (1).

Parameter ε_y is assessed as ~0.10 [42] and the Burgers vector is $b \approx 3\times10^{-10}$ m [9]. Then for different polymers, λ_d assessed by the equation (6) is about 2.5 Å. It is the distance passed by a segment at a shift moving it to the position, shown in Figure 5b, which can be simply calculated from configuration. Thus this assessment also indicates no reasons for assuming any sufficient free path length of dislocations in polymers rather than transition of a segment (or several segments) of macromolecule from one quasi-equilibrium state to another [43].

It is common knowledge [3, 38] that for crystalline materials the Bale-Hirch relation between shear stress, τ_y, and dislocation density, ρ_d, is fulfilled:

$$\tau_y = \tau_{ini} + \alpha G b \rho_d^{1/2} \tag{7}$$

where τ_{ini} is the initial internal stress; α is the efficiency parameter.

Equation (7) is also true for amorphous metals [34]. It was used [33] for describing mechanical behavior of polymers on the example of their main classes. For this purpose, data for amorphous glassy PAr [44], amorphous-crystalline HDPE [45] and epoxy polymer networks (EP) of amine and anhydride cross-linking types [30] were used. Different load conditions were used: axial tension of film samples [44], high-speed bending [45], and axial compression [30]. Figure 8 shows the relation between $Gb\rho_d^{1/2}$ and experimental τ_y values for mentioned polymers, corresponded to equation (7). Clearly they are linear and pass through the origin of coordinates (e.g. $\tau_{ini} = 0$), but α values for linear polymers and polymer networks are different. Thus in the framework of the defect concept suggested, the Baily-Hirch relation is also true for polymers. This indicates that dislocation analogues are correct for any linear defect distorting the ideal structure of the material and creating the field of elastic stresses [33]. From this point of view high defect degree of the polymer structure shall be noted: $\rho_d \approx 10^{10}$ cm^{-2} for crystalline metals [3], $\rho_d \approx 10^{9} \div 10^{14}$ cm^{-2} for amorphous metals [34], and $\rho_d \approx 10^{14}$ cm^{-2} for polymers [41].

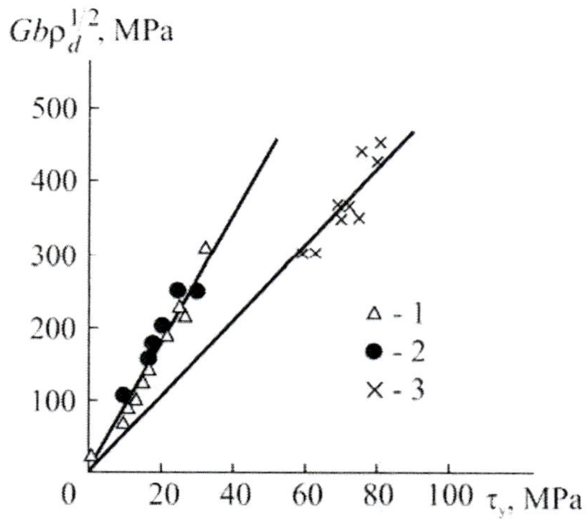

Figure 8. Correlation between $Gb\rho_d^{1/2}$ and flow shear stress τ_y by equation (7) for PAr (1), HDPE (2), and EP (3) [33]

As a consequence, the results obtained [33] indicate that in contrast with metals, for polymers realization of the Frenkel mechanism during yielding is much more probable rather than the motion of defects (Figure 5). This is stipulated by the above-discussed principal (even strictly antipodal) differences in the structure of crystalline metals and polymers [2].

Let us consider the assessment methods for Burgers vector, b. It is suggested that for polymers the value of b must be corresponded to their molecular characteristics. For example, the following formula was deduced [9]:

$$b = \frac{4}{3\pi}\left(r_{ch} + l_m\right) \tag{8}$$

where r_{ch} is the distance between neighboring chains; l_m in the monomeric unit length.

One more simple technique for b assessment has been suggested [46]. It is known that there is a linear correlation between the shear modulus G and the glass transition temperature T_g (Figure 9). In the framework of the cluster model, theoretical correlation between G and T_g is as follows [46]:

$$\frac{G}{T_g} = \frac{4\pi(1-\nu)k\ln(1/f_c)}{C_\infty b^2 l_0 \ln(r/r_0)} \tag{9}$$

where k is the Boltzmann constant; r and r_0 are external and internal radii of the dislocation force field, respectively. The value r/r_0 is usually accepted equal 10 [47].

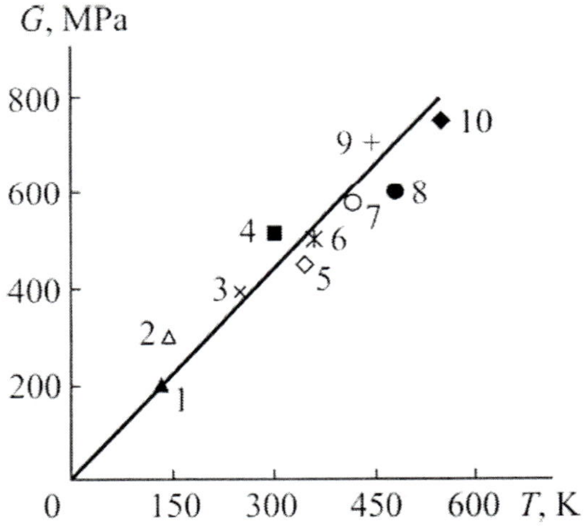

Figure 9. Correlation between shear modulus G and glass transition temperature T_g for poly(tetrafluoroethylene) (1), HDPE (2), PP (3), polyamide-6 (4), PETP (5), PVC (6), PC (7) poly(arylate sulfone) (8), PSF (9), and PAr (10). Values of G were measured for 293 K [46]

Substitution of parameters typical of polymers into equation (9) ($\nu = 0.3$ [47], $f_f = 0.07$ [47], $C_\infty = 4.0$ [48], $b = 4$ Å [9], $l_0 = 1.54$ Å [48]) gives $G/T_g = 1.40$ MPa/K, which is close to $G/T_g = 1.50$ MPa/K, obtained from data in Figure 9. Parameters C_∞ and b in equation (9) are the most sensitive to the origin of polymers. The rest characteristics can be accepted practically constant for the majority of polymers. It follows from the condition G/T_g = const and equation (9) that Burgers square vector is in inverse proportion to characteristic relation C_∞. If the condition $G/T_g = 1.50$ MPa/K and the following constants in equation (9) are used: $\nu = 0.30$; $f_f = 0.07$; $l_0 = 1.54$ Å; $r/r_0 = 10$, then for b assessment the following empirical equation can be obtained [46]:

$$b = \left(\frac{60.7}{C_\infty}\right)^{1/2}, \text{Å}. \tag{10}$$

As a consequence, an increase of macromolecule flexibility (C_∞ increase) causes a decrease of the Burgers vector (a decrease of the distance between chains). As C_∞ varies within the range of 2.2 – 10.0 [48], calculation by equation (10) gives b = 2.3 – 4.9 Å which is in good correspondence with the known data on polymers [8, 9]. Table 1 shows comparison of b values estimated by the mentioned method and the ones obtained from the literature for five polymers. Clearly good conformity is observed again [46].

The Poisson coefficient is one of the most important characteristics of polymers, in many ways determining their mechanical and heat-physical properties [50, 51]. Naturally, a tight interconnection of this parameter with the polymer structure is suggested; however, such correlation has yet been obtained on the basis of the above-discussed structural defect concept [52].

Table 1. Comparison of the Burgers vector b values calculated by equation (10) and obtained from the literature [49]

Polymer	b, Å		
	Reference [8]	Reference [9]*	Equation (3.10)**
Polystyrene	4.5	3.1 (2.8)	2.46
Poly(methyl methacrylate)	4.3	2.9 (2.7)	2.94
Polyvinylchloride	5.8	2.3 (2.5)	3.0
Poly(ethylene terephthalate)	6.2	-	3.80
Polycarbonate	5.6	-	5.03

* Values in brackets are experimental [9].

** Calculation of b by equation (10) applied C_∞ values, shown in ref. [48].

As is known [53], one of the laws in the basis of the theory of continuous medium elasticity is experimentally determined Poisson relation postulating the effect of transverse deformations, $\varepsilon_\perp$:

$$\varepsilon_\perp = \nu\varepsilon_\|, \tag{11}$$

where ν is the Poisson coefficient; $\varepsilon_\|$ is the longitudinal deformation.

One of the methods for quantitative estimation of ν is calculation of it from experimental values of elasticity modulus, E, and the strain yield, σ_y [51]:

$$\frac{\sigma_y}{E} = \frac{(1-2\nu)}{6(1+\nu)} \tag{12}$$

The equation (12) represents quite strict correlation, deduced from the polymer state equation [51]. However, it does not indicate the connection between ν and the structure. This connection can be obtained from consideration of the local order zone as a deviation from the ideal structure; hence; the segments in it are identified as linear defects. In such an interpretation using mathematical apparatus of the theory of dislocations, one can obtain the following expression [54]:

$$\frac{\sigma_y}{\sqrt{3}} = \frac{Eb\rho_d^{1/2}}{4\pi(1+\nu)} \tag{13}$$

Combination of equations (8), (1), (10), and (13) gives the following expression for determination of ν by structural parameters of polymer only [49]:

$$\nu = 0.5 - 3.22\times10^{-10}(l_0\,\nu_{cl})^{1/2}. \tag{14}$$

Figure 10 shows comparison of the theoretical temperature dependence of ν (equation (14)) and its values estimated from equation (12) for blow and quasi-static tests. Expectedly

[55, 56], ν increases with T for all three data series. Maximal deviation between theoretical ν values and appropriate blow test results does not exceed 20%, and for quasi-static tension it is even lower – below 10% [49].

Obviously the above-shown data indicate that calculation of the Poisson coefficient by structural characteristics of the polymer only gives good correspondence with results obtained by other methods. It should be noted also that equation (14), used for theoretical assessment of ν, includes structural characteristics only in the absence of any adjustable parameter [49].

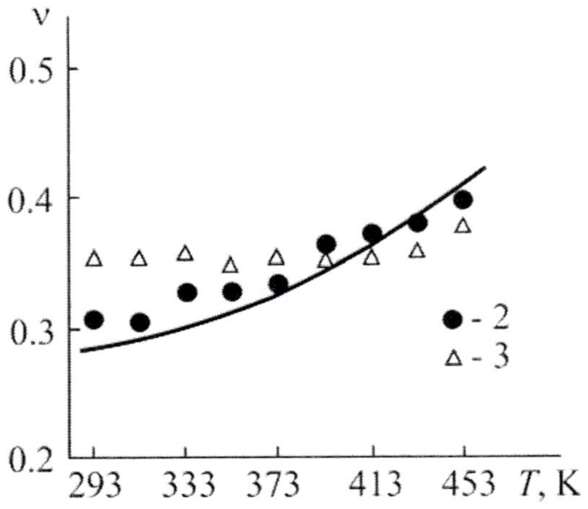

Figure 10. Temperature dependence of the Poisson coefficient ν for poly(arylate sulfone): 1 – calculation by equation (14); 2 and 3 – calculation by equation (12) for quasi-static (2) and blow (3) tests [49]

Completing the Chapter, let us discuss an example of "classic" dislocation analogy modification for polymers using the suggested concept of the structural defect. The dislocation model by Crist [57] suggests formation of a spiral dislocation (or a couple of such dislocations) with Burgers vector b in crystalline zones of polymers, and yielding is realized during formation of a critical germinal zone of u^* dimension:

$$u^* = \frac{K_{el} b}{2\pi \tau_y} \tag{15}$$

where K_{el} is an elastic constant; τ_y is the shear strain yield.

In its turn, the zone sized u^* is formed at overcoming the energetic barrier ΔG^* [57]:

$$\Delta G^* = \frac{K_{el} b^2 l_c}{2\pi} \left[\ln\left(\frac{u^*}{r_0} \right) - 1 \right] \tag{16}$$

where l_c is the dislocation length equal the crystallite length; r_0 is the dislocation center radius.

The equation (15) allows theoretical assessment of τ_y in the presence of the known parameters u^*, K_{el} and b. Let us consider their determination in relation to the structure of amorphous glassy polymers.

As suggested in the simulation [57], the germinal zone sized u^* is formed in a defect-free area of amorphous-crystalline polymer, i.e. in a crystallite. In the framework of the concept suggested for the structural defect and applied to the structure of amorphous glassy polymer, this area is represented by a packless matrix surrounding a local order zone (cluster). The structure of the latter zone is quite similar to the defect-free polymer, postulated by the Flory "felt" model [16, 17]. It has been shown [31] that during the flow process this structural component is subject to mechanical devitrification. In this interpretation, u^* can be determined as follows [58]:

$$u^* = R_{cl} - r_{cl}, \tag{17}$$

where R_{cl} is the half distance between centers of neighbor clusters; r_{cl} is the cluster radius.

For PC and PAr, R_{cl} values are shown in the work [59] (they can also be calculated by the formula (14)); r_{cl} can be determined from the relation as follows [60]:

$$r_{cl} = \left(\frac{n_{cl} S}{\pi \eta} \right)^{1/2} \tag{18}$$

where n_{cl} is the number of segments in the cluster: S is the cross-section of the macromolecule; η is the packing factor equal 0.868 (for close packing) [60].

For K the shear modulus, G, must be accepted; b value is determined from the equation (10). Similar to the previous, n_{cl} equals $1/2F$. Thus as follows from the above-said, the basic role in τ_y or σ_y determination (because $\sigma_y = \sqrt{3}\,\tau_y$) is played by structural (R_{cl} and F) and molecular (S and C_∞) characteristics of the polymer.

Figure 11 shows comparison of temperature dependencies of σ_y for PC and PAr, experimental and calculated by the method described. Clearly good conformity between theory and experiment was obtained. Note that adjustable parameters were not also used for σ_y calculation.

Figure 12 presents comparison of $\sigma_y(T)$ dependencies for HDPE, experimental and calculated by the equation (10). In this case, no conformity similar to the one observed between PC and PAr (Figure 11) was obtained. This is simply explained by the application of HDPE structure only to calculation of its σ_y. That is why σ_y values are much lower than the experimental ones. Nevertheless, of the extreme importance is that for amorphous-crystalline polymers (polyethylenes, in particular), σ_y is fully determined by the contribution of crystalline zones [57, 61 – 63]. The latter can be calculated with regard to technique, described in the work [54], and the totality of contributions of both structural components of HDPE into the strain yield is also presented in Figure 12 as the function of test temperature. In the latter case, good conformity between theory and experiment is obtained [58].

Thus following the interpretation of the ideal structure and defect for the polymer amorphous state, presented in this Chapter, good correspondence of theory to experimental results for both amorphous glassy and amorphous-crystalline polymers was obtained [58]. In this case, molecular and structural characteristics of polymers were used as the initial data. Hence, by virtue of the mentioned reciprocal of ideal structures for crystalline solids and polymer amorphous state, the yielding mechanisms for them is generation and motion of linear defects in the long range order zones and annihilation of the same defects in disordered zones, respectively.

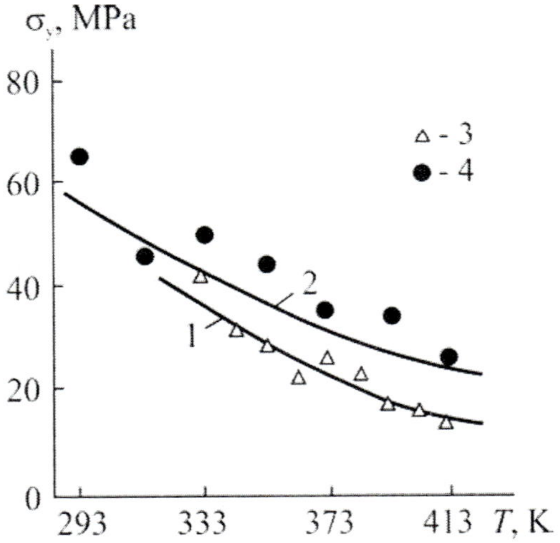

Figure 11. Temperature dependencies of σ_y: (1, 2) experimental and (3, 4) calculated by equation (15) for PC (1, 3) and PAr (2, 4) [58]

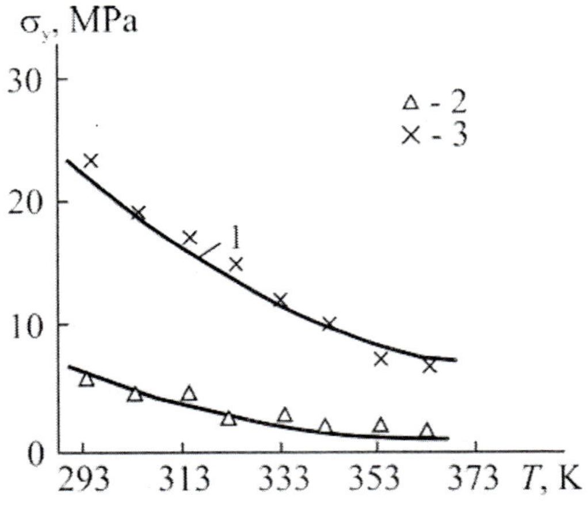

Figure 12. Temperature dependencies for HDPE: (1) experimental and (2, 3) theoretical; 2 – calculated by equation (15), 3 represents the totality of amorphous and crystalline phases contributions into σ_y [58]

Further on, let us discuss the assessment of the energetic barrier, ΔG^*, in the framework of the cluster model. As shown [47], annihilation (transition into the packless matrix) of a segment participating in the cluster requires energy U_{cl}:

$$U_{cl} = \frac{Gb^2 l_0 C_\infty}{4\pi(1-\nu)} \ln\left(\frac{r}{r_0}\right). \quad (19)$$

Figure 13 shows comparison of ΔG^* values, calculated from the equation (16), where for l_s the length of statistical segment (l_{st}) is taken, and U_{cl} ones, calculated by the equation (19), for PC and PAr. These data indicate good quantitative conformity of ΔG^* и U_{cl} parameters. This proves the conclusion, made in the framework of the model [2] that energy U_{cl}, required for segment separation from the cluster (e.g. annihilation of a linear defect), represents the energetic barrier for the flow process of amorphous glassy polymers.

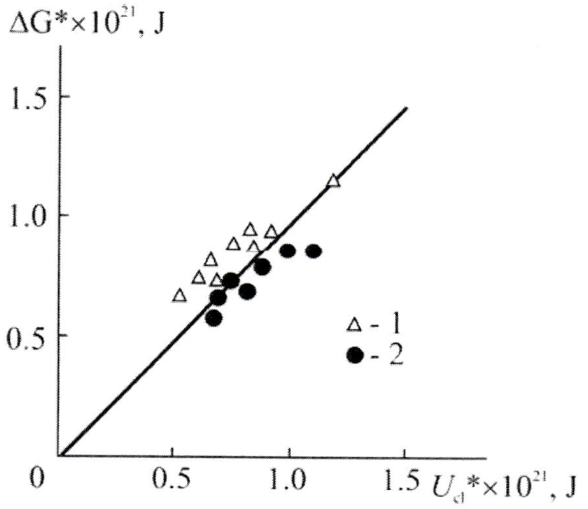

Figure 13. Correlation between critical energy of the germinal zone formation under yielding, ΔG^*, and linear defect annihilation energy, U_{cl}, for PC (1) and PAr (2) [58]

The interpretation suggested in this Chapter is in good conformity with other models, developed for description of polymer yielding. The equation for assessment of critical radius of the dislocation loop, R_l, is shown below [6]:

$$R_l = \frac{Gb}{4\pi\tau_y}\left[\ln\left(\frac{2R_l}{b\sqrt{3}}\right)+2\right]. \quad (20)$$

Calculations performed by equation (20) give R_l equal 38.7 Å for PC and 26 Å for PAr (at T = 293 K). These values are in good conformity with appropriate R_{cl} values, obtained in the framework of the cluster model (31.1 Å and 25 Å, respectively [59]).

In the work [64] the flow process in amorphous glassy PC is associated with generation of quasi-point defects with increased molecular mobility. Obviously in the framework of the current model, mechanically devitrified germinal zone sized u^* represents the analogue for the mentioned defect.

Thus the results obtained [58] have proved once again correctness of the cluster model of the amorphous polymer state structure, interpretation of linear defect in this structure and the yield mechanism applied to description of amorphous glassy and amorphous-crystalline polymer yielding. Of special attention is the existence of significant (about 25%) contribution of the amorphous phase in amorphous-crystalline polymers into their strain yield, σ_y [58].

REFERENCES

[1] Kozlov G.V., Beloshenko V.A., and Varyukhin V.N., *Applied Mechanics and Technical Physics*, 1996, vol. 37(3), pp. 115 – 119. (Rus)

[2] Kozlov G.V., Belousov V.N., Serdyuk V.D., and Kuznetsov E.N., *High Pressure Physics and Technology*, 1995, vol. 5(3), pp. 59 – 64. (Rus)

[3] Honeycombe R.W.K., *The Plastic Deformation of Metals*, London, Edward Arnold Publishers Ltd., 1968, 497 p.

[4] Argon A.S., *Phil. Mag.*, 1973, vol. 28(4), pp. 839 - 865.

[5] Argon A.S., *J. Macromol. Sci.-Phys.*, 1973, vol. B8(3-4), pp. 573 - 596.

[6] Bowden P.B. and Raha S., *Phil. Mag.*, 1974, vol. 29(1), pp. 148 - 165.

[7] Escaig B., *Ann. Phys.*, 1978, vol. 3(2), pp. 207 - 220.

[8] Pechhold W.R. and Stoll B., *Polymer Bull.*, 1982, vol. 7(4), pp. 413 - 416.

[9] Sinani A.B. and Stepanov V.A., *Mechanics of Composite Materials*, 1981, vol. 17(1), pp. 109 – 115. (Rus)

[10] Oleinik E.F., Rudnev S.N., Salamatina O.B., Nazarenko S.I., and Grigoryan G.A., *Doklady AN SSSR*, 1986, vol. 286(1), pp. 135 – 138. (Rus)

[11] Melot D., Escaig B., Lefebvre J.M., Eustache R.P., and Laupretre F., *J. Polymer Sci.: Part B: Polymer Phys.*, 1994, vol. 32(11), pp. 1805 - 1811.

[12] Boyer R.F., *J. Macromol. Sci.-Phys.*, 1976, vol. B12(12), pp. 253 – 301.

[13] Fischer E.W. and Dettenmaier M., *J. Non-Cryst. Solids*, 1978, vol. 31(1-2), pp. 181 - 205.

[14] Wendorff J.H., *Polymer*, 1982, vol. 23(4), pp. 543 - 557.

[15] Nikol'sky V.G., Plate I.V., Fazlyiev F.A., Fedorova E.A., Filippov V.V., and Yudaeva L.V., *Vysokomol. Soedin.*, 1983, vol. A25(11), pp. 2366 – 2371. (Rus)

[16] Flory P.J., *Pure and Appl. Chem.*, 1984, vol. 56(3), pp. 305 - 312.

[17] Flory P.J., *Brit. Polymer J.*, 1976, vol. 8(1), pp. 1 - 10.

[18] Kozlov G.V., Kuznetsov E.N., Beloshenko V.A., and Lipatov Yu.S., *Doklady NAN Ukrainy*, 1995, No. 11, pp. 102 – 104. (Rus)

[19] Lin C., Busse L.E., Nagel S.R., and Faber J., *Phys. Rev. V.*, 1984, vol. 29(9), pp. 5060 - 5062.

[20] Kozlov G.V., Beloshenko V.A., and Lipskaya, *Ukrainsky Fizichesky Zhurnal*, 1996, vo. 41(2), pp. 222 – 225. (Rus)

[21] Kozlov G.V. and Zaikov G.E., *Izv. KBNTs RAI*, 2003, No. 9, pp. 132 – 140. (Rus)

[22] Ginier A., *X-Ray Diffraction Analysis of Crystals*, Moscow, Fizmatgiz, 1961, 604 p. (Rus)

[23] Majid C.A., Ahmad A.U., Akber R.A., and Khan H.A., *Radiat. Phys. Chem.*, 1980, vol. 16(5), pp. 379 - 383.

[24] Petrie S.E.B., *J. Macromol. Sci.-Phys.*, 1976, vol. B12(2), pp. 225 - 247.

[25] Kozlov G.V., Sanditov D.S., and Lipatov Yu.S., In Coll.: *Fractals and Local Order in Polymeric Materials*, Ed. G.V. Kozlov and G.E. Zaikov, Huntington, New York, Nova Science Publishers, 2001, pp. 65 - 82.

[26] Ball R.C., *Physica D.*, 1989, vol. 38(1), pp. 13 - 15.

[27] Song H.-H. and Roe R.-J., *Macromolecules*, 1987, vol. 20(11), pp. 2723 - 2732.

[28] Murthy N.S., Minor H., Bednarczyk C., and Krimm S., *Macromolecules*, 1993, vol. 26(7), pp. 1712 - 1721.
[29] Kumar S. and Adams W.W., *Polymer*, 1987, vol. 28(8), pp. 1497 - 1504.
[30] Beloshenko V.A. and Kozlov G.V., *Mechanics of Composite Materials*, 1994, vol. 30(4), pp. 451 – 454. (Rus)
[31] Kozlov G.V., Beloshenko V.A., Gazaev M.A., and Novikov V.U., *Mechanics of Composite Materials*, 1996, vol. 32(2), pp. 270 – 278. (Rus)
[32] Kozlov G.V., Shogenov V.N., and Mikitaev A.K., *Doklady AN SSSR*, 1988, vol. 298(1), pp. 142 – 144. (Rus)
[33] Kozlov G.V., Afaunova Z.I., and Zaikov G.E., 'Theoretical assessment of polymer yielding', *Electron J. "Studied in Russia"*, 2002, vol. 98, pp. 1071 – 1080; *http: //zhurnal.ape.relarn.ru/articles/2002/098.pdf//.* (Rus)
[34] Liu R.S. and Li J.Y., *Mater. Sci. Eng.*, 1989, vol. A114(1), pp. 127 - 132.
[35] Aleksanyan G.G., Berlin Al.Al., Gol'dansky A.V., Grineva N.S., Onishchuk V.A., Shantarovich V.P., and Safonov G.P., *Khimicheskaya Fizika*, 1986, vol. 5(9), pp. 1225 – 1234. (Rus)
[36] Balankin A.S., Bugrimov A.L., Kozlov G.V., Mikitaev A.K., and Sanditov D.S., *Doklady Akademii Nauk*, 1992, vol. 236(3), pp. 463 – 466. (Rus)
[37] Kozlov G.V., and Novikov V.U., *Uspekhi Fizicheskikh Nauk*, 2001, vol. 171(7) pp. 717 – 764. (Rus)
[38] McClintock F.A. and Argon A.S., *Mechanical Behavior of Materials*, Addison-Wesley Publishing Company, Inc., Reading, Massachusetts, 1966, 514 p.
[39] Wu S., *J. Appl. Polymer Sci.*, 1992, vol. 46(4), pp. 619 - 624.
[40] Sanditov D.S. and Bartenev G.M., *Physical Properties of Irregular Structures*, 1982, Novosibirsk, Nauka, 256 p. (Rus)
[41] Milman L.D. and Kozlov G.V., In Coll.: *Polycondensation Processes and Polymers*, Ed. A.K. Mikitaev, 1986, Nalchik, KBGU, pp. 130 – 141. (Rus)
[42] Shogenov V.N., Kozlov G.V., and Mikitaev A.K., *Vysokomol. Soedin.*, 1989, vol. A31(8), pp. 1766 – 1770. (Rus)
[43] Peschanskaya N.N., Bernstein V.A., and Stepanov A.A., *Fizika Tverdogo Tela*, 1978, vol. 20(11), pp. 3371 – 3374. (Rus)
[44] Shogenov V.N., Belousov V.N., Potapov V.V., Kozlov G.V., and Prut E.V., *Vysokomol. Soedin.*, 1991, vol. A33(1), pp. 155 – 160. (Rus)
[45] Mashukov N.I., Gladyishev G.P., and Kozlov G.V., *Vysokomol. Soedin.*, 1991, vol. A33(12), pp. 2538 – 2546. (Rus)
[46] Sanditov D.S. and Kozlov G.V., *Fizika I Khimia Stekla*, 1993, vol. 19(4), pp. 593 – 601. (Rus)
[47] Kozlov G.V. and Sanditov D.S., *Anharmonic Effects and Physicomechanical Properties of Polymers*, 1994, Novosibirsk, Nauka, 261 p. (Rus)
[48] Aharoni S.M., *Macromolecules*, 1983, vol. 16(9), pp. 1722 - 1728.
[49] Kozlov G.V., Belousov V.N., Serdyuk V.D., Mikitaev A.K., and Mashukov N.I., *Doklady Adyigeiskoi (Cherkesskoi) Mezhdunarodnoi AN*, 1997, vol. 2(2), pp. 88 – 93. (Rus)
[50] Balankin A.S., *Synergism of Deformable Body*, 1991, Moscow, MO SSSR, 404 p. (Rus)
[51] Novikov V.U. and Kozlov G.V., *Analysis of Polymer Degradation in the Framework of the Fractal Concept*, 2001, Moscow, Izd. MGOU, 135 p. (Rus)

[52] Kozlov G.V., Belousov V.N., Sanditov D.S., and Mikitaev A.K., *Izv. VUZov, Severo-Kavkazskii Region. Estestvennye Nauki*, 1994, No. 1-2(86), pp. 54 – 58. (Rus)

[53] Balankin A.S., *Doklady AN SSSR*, 1991, vol. 319(5), pp. 1098 – 1101. (Rus)

[54] Belousov V.N., Kozlov G.V., Mashukov N.I., and Lipatov Yu.S., *Doklady RAN*, 1993, vol. 328(6), pp. 706 – 708. (Rus)

[55] Filyanov E.M., *Vysokomol. Soedin.*, 1987, vol. A29(5), pp. 975 – 981. (Rus)

[56] Hohg S.-D., Chung S.Y., Fedors R.F., and Moacanin J., *J. Polymer Sci.: Polymer Phys. Ed.*, 1983, vol. 21(9), pp. 1647 - 1660.

[57] Crist B., *Polymer Commun.*, 1989, vol. 30(3), pp. 69 - 71.

[58] Kozlov G.V., Zaikov G.E., Burmistr M.V., and Korenyako V.A., *Voprosy Khimii i Khimicheskoi Tekhnologii*, 2002, No. 6, pp. 81 – 84. (Rus)

[59] Kozlov G.V., Belousov V.N., and Mikitaev A.K., *High Pressure Physics and Technology*, 1998, vol. 8(1), pp. 64 – 70. (Rus)

[60] Kozlov G.V., Shogenov V.N., and Mikitaev A.K., *Inzhenerno-Fizicheskii Zhurnal*, 1988, vol. 71(6), pp. 1012 – 1015. (Rus)

[61] Popli R. and Mandelkern L., *J. Polymer Sci.: Part B: Polymer Phys.*, 1987, vol. 25(3), pp. 441 - 483.

[62] Bernstein V.A., Sirota A.G., Egorova L.M., and Egorova V.M., *Vysokomol. Soedin.*, 1989, vol. A31(4), pp. 776 – 779. (Rus)

[63] Kennedy M.A., Peacock A.J., and Mandelkern L., *Macromolecules*, 1994, vol. 27(19), pp. 5297 - 5310.

[64] G'Sell C., Bari H.E., Perez J., Cavaille J.Y., and Johari G.P., *Mater. Sci. Eng.*, 1989, vol. A110(2), pp. 223 - 229.

In: Frontiers in Physical Organic Chemistry
Editor: G.E. Zaikov, pp. 103-156

ISBN 1-60021-128-3

Chapter 7

SPECIFIC AND NONSPECIFIC INTERACTIONS IN COMPLEX FORMING REACTIONS

R. G. Makitra, A. A. Turovsky and G. E. Zaikov*

L.V.Pisarzhevskii Institute of Physical Chemistry, Ukranium National Academy of Sciences, 3A, Naukova st., Lviv – 79053, Ukraine

*N.M.Emanuel Institute of Biochemical Physics Russian Academy of Sciences, 4, Kosygin St., Moscow 119991, Russia

Besides data generalization for the environment effect on chemical reaction kinetics, empirical solvent "polarity" scales are very widely used for spectral events. However, due to insufficient universality of these scales, to set the connection between solvent properties and spectral shifts of substances solved in the solvents, expressions considering several parameters were used for the first time. Besides the known Bakshiev and Mac-Ray formulae, the work by Fowler and Katritzky [1] should be mentioned. In this work the solvent effects on spectral and kinetic processes are described using multiparameter, linear equations. Correlated parameters in these equations are various functions of light refraction and dielectric constant.

In 1972, I.A. Koppel and V.A. Palm [2] suggested a four-parameter equation taking into account both nonspecific and specific solvation. It was found highly effective for setting a connection between the environment properties, kinetic and spectral parameters of solutes. It also shows [2] examples of positive generalization of data in more than 30 UV-, 12 IR- and 5 NMR reaction series. Since that time some works were published, showing numerous examples of application of the Koppel-Palm formula or other multiparameter equations (Kryigovsky and Fawcett, Kamlet-Taft, etc.) to generalization of experimental data on the environment effect on solute spectral parameters.

Finally, shifts in spectra of solutes are stipulated by the interaction between the environment and solute which, in the limit, leads to formation of compounds of a definite composition - complexes. Thus the solvent is simultaneously the environment and the reagent. It is known that spectral methods for studying charge transfer complexes (Benesi-Hildebrand, Ketelaare, etc.) are based on the study of spectra for binary mixtures of different donor and acceptor ratios in a relatively inert solvent (hydrocarbons, halohydrocarbons). Relatively low solvation ability of the solvent will induce predominance of deficient component selective solvation by the second, excessive component in the system. Thus is may be expected that multiparameter equations will also be found useful for setting a connection between properties of excessive component (which is usually a donor) and thermodynamic parameters of the complex formation (equilibrium constants and complex formation enthalpies). Actually, it is found that multiparameter equations are useful for detecting both a connection between solvent properties and spectral parameters of solutes and dependence between properties of excessive component and spectral characteristics or thermodynamic indices of a charge transfer complex, formed with an acceptor. It is also suitable for taking into account the effect of environment properties on thermodynamic indices for generalizing results on shift enthalpies in binary systems, principally leading to the same complex formation.

1. The Environment Effect on Complexes with Hydrogen Bond and Charge Transfer

1.1. Linear Free Energy (LFE) Principle Application to Generalization of Data on Organic Complex Formation

Since 1950's the problem of acid-basic properties of organic substances began intensively developing. For instance, many works were devoted to an interrelation between complex formation (interaction) enthalpies, determined using spectral or calorimetric techniques, and basicity of organic donors. Gutmann [3] suggested a scale of organic substance basicity values, the so-called donor numbers *DN*, numerically equal to enthalpies of donor mixing with $SbCl_5$ in dichloroethane. It was also suggested that this scale might characterize spectrally determined enthalpies of complex formation of the same donors with iodine, phenol, $SbCl_3$, Me_3SnCl, enthalpies of ion transfer from one medium to another (i.e. their re-solvation enthalpies), etc. However, later studies show that such proportionality is kept only approximately and sometimes is associated with Gutmann's consideration of deficient number of solvents. Drago [4, pp. 104 – 105] indicated the absence of a reliable relation between ΔH_{J_2} and ΔH_{PhOH}, as well as between these values and *DN*. A satisfactory proportionality was not also observed between calorimetric-determined enthalpies of donor mixing with $AsCl_3$, $SbCl_3$, $AlBr_3$, and Br_2 and *DN*:

$$-\Delta H_{SbCl_3} = (-1.95 \pm 3.9) + (0.55 \pm 0.15)DN;\ n = 11;\ R = 0.78; \quad (1)$$

$$-\Delta H_{AsCl_3} = (-1.68 \pm 1.53) + (0.26 \pm 0.06)DN;\ n = 9;\ R = 0.85; \quad (2)$$

$$-\Delta H_{Br_2} = (1.56 \pm 1.45) + (0.65 \pm 0.07)DN;\ n = 7;\ R = 0.93; \quad (3)$$

Moreover, *DN* values (as well as general values of substance interaction enthalpies) depend on the solvent used, where these values are determined.

Therefore, there is a tendency to characterize interaction enthalpies by semi-empiric multiparameter equations. Note that Gutmann concluded soon that two scales should be used – basicity (donor numbers *DN*) and electrophilicity (acceptor numbers *AN*)of the solvents, and the application of one or another scale depends on the process type. The acceptor number scale is approximately proportional to previously suggested solvent "polarity" scales – Z by Kosower, Y by Grunwald-Winstein, and E_T by Dimroth-Reichardt [5].

A.V. Iohansen showed that hydrogen bond formation and spectral band intensity and A-H bond absorption maximum shift enthalpies at hydrogen bond formation are satisfactorily described by the so-called factor rule:

$$\Delta H_{ij}(\Delta I_{ij}, \Delta \nu_{ij}) = \text{const}\cdot P_i E_j, \quad (4)$$

where P_i and E_j are electronegativity and electron-donor factors of interacting substances [6]. P and E values are obtained basing on taking average on a data series on enthalpies and other parameters of the hydrogen bond formation. For the scale setting, it is postulated that $P = 1$ for phenol and $E = 1$ for diethyl ester. As a consequence, const equals enthalpy of hydrogen bond formation between phenol and diethyl ester, which is 5.3 kcal/mol. The Iohansen factor rule satisfactorily converge with the experiment, but it can be used until complexes with hydrogen bonds are formed.

To characterize donor-acceptor interactions, first of all, interaction enthalpies, Drago *et al.* [4, 7] suggested a four-parameter equation:

$$\Delta H = E_A E_D + C_A C_D, \quad (5)$$

where parameters E and C are determined empirically, by generalizing a broad experimental data array on enthalpies of complex formation with iodine and phenol, and correspond to ability of donor D and acceptor A to "covalent" and "electrostatic" interaction. These values are associated with total polarization and dipole moment of substances. The use of the above-shown equation allows calculation of ΔH value for unstudied systems. In the case of hydrogen bond and systems with SO_2, standard deviation equals 0.1 – 0.3 kcal/mol, and for higher interaction energy systems it is 0.7 kcal/mol. Unfortunately, in some cases deviations are much higher. For example, calculated enthalpy of BMe_3 interaction with Me_3N equals 25.5 kcal/mol; the same experimental value is 17.6 kcal/mol, and for Me_3SnCl interaction with Et_2O, these enthalpies are 5.6 and 2.2kcal/mol, respectively. Deviations for numerous systems with BF_3 are very much so high. Such deviations are explained by possible steric hindrances and spatial regrouping of substances at complex formation. Solvents may also cause a noticeable effect on ΔH value and correspondingly E and C values that later on induced revision of many of them [4].

Such equations were suggested by other authors (Pearson [8]) basing on the theory of hard and soft bases and acids:

$$\lg K = SASB + \sigma_A \sigma_B, \quad (6)$$

where S is the hardness parameter; σ is the softness parameter. However, Drago excepts against equating these scales.

The works by V.A. Terentiev are based on the same principle. According to his "additive-multiplicative" relation, charge transfer complexes and hydrogen bonds may be described by two parameters characterizing donor and acceptor ability to interaction:

$I = a \times d.$

Therefore, separate characteristics are suggested for enthalpies, equilibrium constants and connection characteristic band shifts:

ΔH_A, K_A, $\Delta \nu_A$ and ΔH_D, K_D, $\Delta \nu_D$, respectively.

The equation $I = a \times d$ is a particular case of full equation for intermolecular interaction:

$$E = \Delta H = (1 - p)ad + pC_AC_D + pI_AI_D, \quad (7)$$

where a and d are dipole components; C_A and C_D are covalent components; I_A and I_D are ionic components; p is polarity of molecules due to charge transfer. If $p \approx 0$, the above-mentioned reduced expression for hydrogen bond energy is obtained. For charge transfer complexes the first term effect is insignificant, and their formation is satisfactorily described by the following expression:

$$\Delta H = pC_AC_D + pI_AI_D, \quad (8)$$

which is close in shape to the Drago equation.

Parameters ΔH_i, K_i and $\Delta \nu_i$ were calculated basing on parameters of "standard" formation of a complex with hydrogen bond from pyridine and phenol in CCl_4:

$\Delta \nu_{D(A)} = 21.10$ cm^{-1}; $\Delta H_{D(A)} = 2.55$ kcal/mol; $K_{D(A)} = 7.8$ mol$^{1/2}$l$^{1/2}$ [9].

In later works, noticeable deviations from the multiplicative rule were observed and its invalidity for equilibrium constant calculation was theoretically proved [10].

The empiricalness of donor and acceptor parameters of substances is general for the above-mentioned works. Krygowski and Fawcett [11] suggested estimation of the acid-base interaction effect using the following two-parameter equation:

$$Q = Q_0 + \alpha A + \beta B, \quad (9)$$

the so-called complementary description to the Lewis acid-base interaction. For acidic property parameter the Reichardt polarity (electrophilicity) scale E_T was taken, and for basic properties - Gutmann's donor numbers *DN*. This equation was suggested for description of changes in some properties of systems (kinetic, spectral and thermodynamic), caused by a solvent. These particular properties are ion activity, solvation enthalpies, standard reduction potentials [11], partial molar dissolution heats of HSO_3F, enthalpies of solute transfer with CCl_4 to HSO_3F, ΔH and ΔG of hydrogen bond formation between fluorophenol and the solvent [11], and mutual dissolution heats for a series of 9 solvents.

However, this expression neglects nonspecific (weak) interactions of the components. Therefore, the dimension of data agreement between calculations and the experiment is frequently low. Generally, the suggested equation is adequate ($R \geq 0.90$) for 75% of the cases under consideration. Therefore, the authors conclude it useful to take into account and additional “entropic” term:

$$Q = Q_0 + \alpha E_T + \beta DN + \gamma^S. \quad (10)$$

Further on, Krygowski and Fawcett used the characteristic vector technique [12] for better outlining the effect of separate factors of specific interaction on dissolution heats of monad ions or organic substances.

The same equation was successfully applied [13] to associate equilibrium constants of complex formation with donor and acceptor properties of the solvents. The only difference was in replacing the polarity parameter E_T by Gutmann’s acceptor number AN. For some complex forming systems, for example, $[Co(en)_2Cl_2]^+ + Cl^-$, $J_2 + J^-$, etc., the following equation:

$$\lg K = \alpha DN + \beta AN + \gamma \quad (11)$$

gave high coefficients of correlation, $R = 0.97 - 0.99$. The coefficient α is accepted as the measure of Lewis’s acidity in relation to standard acceptor $SbCl_5$.

According to model ideas, developed by Mayer [14], variation of isobaric thermodynamic potential in the system during dissolution is defined by the equation:

$$\Delta G_{solv(inv)} = \Delta G_{cav} + \Delta G_{conc} + \Delta G_{sp} + \Delta G_{dp}, \quad (12)$$

where the appropriate terms are energy effects of cavity formation in the environment ΔG_{cav}, nonspecific, concentration-dependent effect ΔG_{conc}, short-range, specific interaction ΔG_{sp}, and the interaction, induced by the environment dielectric polarization ΔG_{dp}.

The change of solvation effects at substance transfer from one environment to another is described by the following equation:

$$\Delta\Delta G_{solv} = a\Delta DN + b\Delta AN + c\Delta\Delta G_{cav}. \quad (13)$$

This equation was found effective for generalizing data on alkaline metal salt solubility, equilibrium constants of $[CoCl_4]^-$, $[AgCl_2]^-$ and $[AgBr_2]^-$ complex formation, and solvent effects on some kinetic parameters.

Based on statistical treatment of large data array for the solvent effect on physiochemical parameters of solutes, new scales of anionic and cationic salvation ability of solvents A and B were suggested [15]. The suggested equation:

$$p_{ij} = a_i A_i + b_j B_j + C \quad (14)$$

is useful also for processing data on spectrum shifts and complex formation equilibriums.

Therefore, several expressions were suggested for characterizing spectral shifts of substances in various solvents, constants of complex formation from solutes with solvents or the solvent effect as the environment on dissolved complexes. However, these expressions usually neglect solvation processes, mostly nonspecific solvation and thus their applicability is limited. As shown below, the adequate interrelation between these parameters and solvent properties may be obtained only if all energy changes happening in the system during dissolution are taken into account. For this purpose, basing on the Mayer approach [14], Koppel-Palm [2] or Kamlet-Taft-Abraham formulae are used. In this case, complexes with hydrogen bonds are considered as a particular case.

1.2. Some Systems with Strong Hydrogen Bond

Hydrogen bonds are still the target of intensive investigations in chemistry. Nevertheless, some problems remain unclear, for example, a possibility to correlate OH frequency shifts with the donor ionization potential or dipole moment of a particle, the boundaries of Badger-Bauer rule validity, etc. [16 – 19]. Bellamy *et al.* [20, 21] and Koppel and Payu [22] indicated satisfactory mutual linearity (R = 0.97 – 0.99) between X-H band shifts in IR-spectra for relatively strong proton donors (phenols, pyrrol, HCl, HNCO, methanol). As a result, $\Delta\nu_{XH}$ values are defined by basicity of proton acceptors

$$B \equiv \Delta\nu_{PhOH} \text{ and } \Delta\nu_{XH}^{CCl_4} = a + b\Delta\nu_{PhOH}^{CCl_4} \equiv a + bB. \quad (15)$$

However, in mentioned [20, 21] and later works some deviations from such proportionality were observed. In particular, *ortho*-nitrophenol [23], carboxylic acids, weak proton donors as $CHCl_3$, etc. do not fit the proportionality. It seems logical that the reason for such deviations is contribution of other effects, for example, competitive formation of intra- and intermolecular hydrogen bonds for *ortho*-nitrophenol and, which is most probable for subacid proton donors of chloroform kind, a significant contribution of other, mostly nonspecific, solvation effects into total interaction. Such problems concerning contribution of nonspecific interaction were developed before [16].

It is shown [22] that $\Delta\nu_{X\text{-}H}$ shifts of compounds possessing O-H or N-H bond are proportional to the Palm basicity for a series of proton acceptors. This is also confirmed in the literature [24] by data on H-N band shifts for HNCS and O-D for CH_3OD:

$$\Delta\nu_{NH} = (71.2 \pm 7.2) + (1.908 \pm 0.035)B; \quad (16)$$

n = 18; R = 0.997; S = 14.5;

$$\Delta\nu_{OD} = (27.2 \pm 3.2) + (0.42 \pm 0.01)B; \quad (17)$$

n = 47; R = 0.981; S = 15.0.

Similar results were obtained for subacid *tert*-butyl and cumyl hydroperoxides [25, 26]:

Me_3COOH:

$$\Delta\nu_{OH} = (2.11 \pm 2.71) + (0.77 \pm 0.02)B; \quad (18)$$

$n = 39$; $R = 0.993$; $S = 10.9$;

$PhMe_2COOH$:

$$\Delta \nu_{OH} = (6.90 \pm 3.7) + (0.728 \pm 0.025)B; \quad (17)$$

$n = 31$; $R = 0.984$; $S = 12.3$.

Later data [25] confirm the statement [22] that angular moment coefficients in $\Delta \nu = a + bB$ type equations are proportional to pK of acid dissociation of these compounds in water. Such linear dependence is observed for alcohols, phenols and hydroperoxides, but values for stronger acids, highly ionized in water (HCl, for example) and very soft C-H acids ($CHCl_3$) do not fit this line.

The values $\nu = 65$ and 62^{-1} [27] for OH band shift (*tert*-butyl hydroperoxide) in the presence of dimethyl- and diethylaniline do not also fit the linear dependence [25]. This allows a suggestion that these values correspond to complex formation by π-electrons of an aromatic ring rather than tertiary atom of nitrogen. The first interaction might give much more remote band in the 3220 cm^{-1} region of IR-spectrum. Actually, experimental check proved the presence of two bands at $\nu = 3496$ cm^{-1} [27] and $\nu = 3231$ cm^{-1} (which correlates with theoretically suggested $\Delta \nu$ at 3227 cm^{-1}) in the spectrum of Me_3COOH-dimethylaniline mixture (dissolved in CCl_4).

It is shown [22] that $\Delta \nu_{CH}$ shifts for methanol and *tert*-butanol are associated with the solvent basicity. Treatments of data from the literature [28 – 30] also confirm this conclusion for ethanol, cyclohexane and some fluorinated alcohols. Since coefficients of correlation of these dependencies fall within the range of 0.98 – 0.99, taking into account chargeability of solvents may not noticeably increase R values. However, for n-butanol (data for 19 solvents [31]) taking into account chargeability significantly increases the correlation:

$$\Delta \nu_{OH} = (4.15 \pm 18.67) + (1.434 \pm 0.136)B; \quad (19)$$

$n = 19$; $R = 0.923$;

$$\Delta \nu_{OH} = -87.0 + 306.5\frac{n^2 - 1}{n^2 + 2} + 0.640B; \quad (20)$$

$n = 19$; $R = 0.991$; $S = 15.5$.

Similar to OH valence oscillation shifts, deformation oscillation shifts [32, 33] (Table 1) are also determined by basicity of substances. However, sensitivity of these oscillations to valence oscillations, aimed to lower frequency region, deformation oscillations are shifted to higher frequency region:

CD_3OH:

$\Delta\delta_{OH} = (2.30 \pm 3.43) + (0.145 \pm 0.014)B;$ (21)

$n = 8; R = 0.985;$

Table 1. Deformation ΔδOH and torsional ΔγOH oscillation shifts of alcohols in various environments [32, 33]

Solvents	$\Delta\delta_{OH}$		$\Delta\gamma_{OH}$, cm^{-1}	
	CD_3OH	CCl_3CD_2OH	*t*-BuOH	*c*-hexanol
Cyclohexane	-	-	-23	-3
CCl_4	0	0	0	0
CS_2	-3	2.5	-	-
CH_2Cl_2	-	-	48	57
C_6H_6	10.5	7	52	53
$MeNO_2$	-	-	170	160
PhCN	35	57.5	233	234
MeCN	40.5	59.5	258	242
i-Pr_2O	-	-	300	314
Bu_2O	-	-	312	325
THF	59.5	87.5	313	307
DMSO	80.5	103.5	372	372
Pyridine	84.5	119.5	-	402
Et_2O	-	-	283	291
Et_3N	-	-	425	423

CCl_3CD_2OH:

$\Delta\delta_{OH} = (5.79 \pm 5.47) + (0.270 \pm 0.022)B;$ (22)

$n = 8; R = 0.980.$

Meanwhile, obtaining of a satisfactory correlation between environment properties in environment-induced shifts of torsional oscillation frequencies requires taking into account nonspecific interaction defined by polarity:

tert-BuOH:

$\Delta\gamma_{OH} = (66.89 \pm 27.02) + (0.719 \pm 0.101)B;$ (23)

$n = 13; R = 0.907;$
cyclohexanol:

$\Delta\gamma_{OH} = (66.68 \pm 25.88) + (0.722 \pm 0,090)B;$ (24)

$n = 14; R = 0.918.$

On the other hand, taking into account polarity (the Kirkwood function) dramatically increases R:

tert-BuOH:

$$\Delta\gamma_{OH} = -126.0 + 527.6\frac{\varepsilon - 1}{2\varepsilon + 1} + 0.675B; \quad (25)$$

$R = 0.979$; $S = 35.4$;

cyclohexanol:

$$\Delta\gamma_{OH} = 85.7 + 430.4\frac{\varepsilon - 1}{2\varepsilon + 1} + 0.700B; \quad (26)$$

$R = 0.966$; $S = 44.3$.

Obviously, consideration of nonspecific solvation effects is necessitated by lower value of the energy effect, introduced to these energy transitions by specific interaction. It is unambiguous that in all cases of hydrogen bond formation, besides specific acid-base interaction, less energetically valuable nonspecific interaction proceeds between the proton donor and the environment. This interaction makes definite corrections to resulting experimental values. In higher acidic, fluorinated alcohols, where hydrogen bond is stronger, the effect of nonspecific solvation is less significant; it is more important for very soft C-H acids.

Consideration of data [34] on the environment effect on the $\Delta\gamma_{OH}$ band shift in propanol, bound in a complex with strong electron donor (α-picoline), also indicated the determining role of the environment basicity:

$$\Delta\nu_{OH} = (-0.52 \pm 1.78) + (0.655 \pm 0.049)B; \quad (27)$$

$n = 12$; $R = 0.973$.

The authors used only weakly basic solvents, such as aromatic hydrocarbons. Meanwhile, contrary to spectral shifts of OH band, equilibrium constants of this complex formation weakly correlate with the environment basicity ($R = 0.604$), and other solvation parameters cause a significant effect on lg$K$ value. Total effect of environment properties on lg$K$ value is satisfactorily described by only a five-parameter equation:

$$\lg K = 0.75 - 3.62\frac{n^2 - 1}{n^2 + 2} - 1.65\frac{\varepsilon - 1}{2\varepsilon + 1} + 14.10\delta^2 - 0.0027B - 0.17E_T; \quad (28)$$

$R = 0.965$; $S = 0.08$.

Obviously, consideration of other solvation factors is necessitated by nonlinearity of the relation between lg$K$ and ΔH, i.e. by non-compliance with the Badger-Bauer rule [17].

Contrary to phenols and alcohols, for carboxylic acids a valuable correlation between solvent properties and $\Delta\nu_{OH}$, observed in the solvent environment, may be set by multiparameter equations only. Spectra of acetic acid were studied in detail in 10 solvents [36]. Acetic acid dimerization constant k_D and dimerization heat ΔH_D in these solvents were also obtained. In these studies, correlation between absorption frequency shifts of the monomeric form of acid and $\Delta\nu_{OH}$ and $\Delta\nu_{C=O}$ is absent. A correlation also is not observed between k_D and ΔH_D and solvent polarity, characterized by $\frac{\varepsilon-1}{2\varepsilon+1}$ value. Therefore, it is concluded that the monomeric form of acid specifically interacts with the solvent.

Table 2 shows $\Delta\gamma_{OH}$ and $\Delta\gamma_{C=O}$ values for acetic acid in 10 solvents [36], determined as the difference of absorption frequencies in the current solvent and n-hexane, and dimerization parameters: -ΔH_D, k_D, lgk_D and ΔS. For OH frequencies, the determining parameter is solvent basicity, determined by nucleophilic solvation of a proton by the solvent. However, strict linearity is not observed here, because the paired coefficient of correlation between $\Delta\gamma_{OH}$ and B just equals 0.912 [35].

Table 2. Spectral characteristics of acetic acid in solvents [36]

Solvent	$\Delta\gamma_{OH}$	$\Delta\gamma_{C=O}$	-ΔH_D, kcal/mol	k_D	lgk_D	ΔS, e.u.
n-Hexane	0	0	11.0	1,931	3.286	-22.5
Cyclohexane	3	2	10.8	1,560	3.199	-22.2
CCl_4	10	5	10.5	1,442	3.159	-21.4
Tetrachloroethylene	11	-	10.2	1,437	3.157	-20.3
Trichloroethane	12	14	9.9	1,139	3.057	-19.8
Chloroform	20	21	8.5	747	2.873	-15.8
o-Dichlorobenzene	23	14	8.1	449	2.652	-15.5
Chlorobenzene	28	15	7.6	353	2.548	-14.3
Dichloroethane	55	15	7.1	154	2.188	-14.2
Benzene	70	16	6.4	112	2.049	-12.5

Meanwhile, the use of multiparameter equations allows obtaining of an adequate correlation between $\Delta\gamma_{OH}$ and solvent properties:

$$\Delta\gamma_{OH} = 2.46 - 128.6\frac{n^2-1}{n^2+2} - 179.6\frac{\varepsilon-1}{2\varepsilon+1} + 1.08\delta^2 + 1.067B + 1.68E; \qquad (29)$$

$R = 0.997$; $S = 3.34$;
$r_{01} = 0.443$; $r_{02} = 0.367$; $r_{03} = 0.660$; $r_{04} = 0.912$; $r_{05} = 0.215$.

Since solvent chargeability and electrophilicity parameters are of low significance, they may be eliminated, which gives the following adequate equation:

$$\Delta\gamma_{OH} = -32.63 - 194.8\frac{\varepsilon - 1}{2\varepsilon + 1} + 1.17\delta^2 + 0.984B; \quad (30)$$

$R = 0.980$; $S = 7.60$.

Thus for CH_3COOH, the oscillation frequency shift value $\Delta\gamma_{OH}$ is defined by the solvent basicity. Of a definite importance also is nonspecific solvation and cohesive energy density of the solvents.

Different situation is observed for C=O band shift. For $\Delta\gamma_{C=O}$, the following equation was btained:

$$\Delta\gamma_{C=O} = -44.83 + 263.5\frac{n^2 - 1}{n^2 + 2} + 54.57\frac{\varepsilon - 1}{2\varepsilon + 1} - 0.46\delta^2 - 0.051B + 6.15E; \quad (31)$$

$R = 0.997$; $S = 1.17$;

$r_{01} = 0.513$; $r_{02} = 0.773$; $r_{03} = 0.793$; $r_{04} = 0.729$; $r_{05} = 0.513$.

Clearly the effect of solvent basicity on $\Delta\gamma_{C=O}$ value is negligibly low, and the solvent effect on this value may be described by the four-parameter equation:

$$\Delta\gamma_{C=O} = -38.37 + 236.5\frac{n^2 - 1}{n^2 + 2} + 54.58\frac{\varepsilon - 1}{2\varepsilon + 1} - 0.46\delta^2 + 5.74E; \quad (32)$$

$R = 0.994$; $S = 1.55$.

Therefore, $\Delta\gamma_{C=O}$ value is defined by electrophilic solvation of acetic acid molecule by carbonyl group by the solvent that weakens this bond. This conclusion correlates with direct observation: maximal shifts of C=O frequencies are observed in solvents with the highest electrophilicity (chloroform and dichloroethane). However, a significant contribution in $\Delta\gamma_{C=O}$ value is made by universal interactions, defined by solvent polarity and chargeability parameters. Meanwhile, negativity of the regression coefficient at cohesive energy density indicates a decrease of acetic acid molecule solvation and $\Delta\gamma_{C=O}$ value with intensification of solvent intermolecular interaction.

Acetic acid dimerization constants were also determined [36]. The absence of correlation between ΔH_D and k_D and dielectric constants of solvents was determined, and complex type of the interaction was outlined. Correlation of the same values with solvent parameters by a five-parameter equation gave the following results:

$$-\Delta H_D = 14.48 - 17.97\frac{n^2 - 1}{n^2 + 2} - 1.46\frac{\varepsilon - 1}{2\varepsilon + 1} + 0.017\delta^2 - 0.062B - 0.46E; \quad (33)$$

$R = 0.999$; $S = 0.048$;

$r_{01} = 0.630$; $r_{02} = 0.639$; $r_{03} = 0.798$; $r_{04} = 0.976$; $r_{05} = 0.575$.

After exclusion of low significant components:

$-\Delta H_D = 10.66 - 0.080B - 0.224E$; $R = 0.990$; $S = 0.34$.

For $\lg k_D$ the following equation was deduced:

$$\lg k_D = 3.281 + 2.018\frac{n^2-1}{n^2+2} + 1.748\frac{\varepsilon-1}{2\varepsilon+1} -$$

$$- 0.014\delta^2 - 0.021B - 0.0036E; \quad (34)$$

$R = 0.999$; $S = 0.027$;

$r_{01} = 0.565$; $r_{02} = 0.534$; $r_{03} = 0.578$; $r_{04} = 0.976$; $r_{05} = 0.555$.

and

$$\lg k = 3.747 + 1.826\frac{\varepsilon-1}{2\varepsilon+1} - 0.014\delta^2 - 0.0197B; \quad (35)$$

$R = 0.994$; $S = 0.066$.

Thus acetic acid dimerization degree increases insignificantly with solvent polarity, but dimerization is impeded by proton nucleophilic solvation of the solvent molecule and, to a lesser extent, by intensification of intermolecular interaction in the solvent (given by δ^2 value).

A satisfactory correlation ($R = 0.974$) is observed between $-\Delta H_D$ and $\lg k_D$ (e.g. ΔG_D) values. Similar linear dependence is observed between ΔH_D and ΔS. The higher dimerization of acetic acid is, the more significant are entropy changes (from -12.5 e.u. for benzene to -22.5 e.u. for *n*-hexane). The more powerful interaction between acetic acid molecules and the solvent is (determined by the solvent basicity), the lower is dimerization and changes in entropy and enthalpy in it.

1.3. Proton Donors with C-H Bond

Contrary to relatively strong OH...O bonds, hydrogen bonds formed by activated C-H groups are much weaker and nonspecific solvation is of high importance for them. As an example, the interaction in chloroform-organic donors systems may be considered. The results of physicochemical analysis of such binary systems confirm formation of CTC in the systems. So far as concerns spectral properties of chloroform-donors systems, numerous works are devoted to this problem. However, quantitative dependencies between solvent properties and spectral characteristics were not determined that may be made using multiparameter equations only.

Studies of ESR spectra for chloroform-donors mixtures testify formation of hydrogen bond in these systems [32]. Oscillation IR-frequencies of C-H bond in chloroform in the environment of ester, dioxane or acetone, which form complexes with $CHCl_3$, shift to lower frequency region [38] due to hydrogen bond formation. However, as C-H band is overlapped by absorption bands of the solvents, measurement results are not reliable enough. This correlates with the data [39] indicating a shift of deuterochloroform valence oscillation band C-D = 2254 cm^{-1} to the lower frequency region in the presence of other donor solvents (aromatic hydrocarbons, ketones). Similar variations caused by hydrogen bond formation are also observed in IR-spectra of other halohydrocarbons [40].

Therefore, these data prove the presence of valence and deformation C-H (or C-D) oscillations in chloroform in donor solvents due to hydrogen bond formation. However, usually systems with 1 or 2 solvents were studied that gave no opportunity to compare shift values in IR-spectrum with the solvent properties. It is just reported [41] that IR-spectrum anomalies in C-H valence oscillations for $CHCl_3$, $CHBr_3$, $CHCl{=}CCl_2$ in solvents are associated with specific interaction C-H-X, because no direct correlation between the oscillation frequency and physical properties of the solvent (dielectric constant, first of all) is observed.

Specific interaction of organohalogen compounds with electron donors affects frequencies of C-Cl bond deformation oscillations. IR-spectra for CH_2ClOCH_3 were studied in several solvents [42]. It is found that C-Cl band shift at ν = 679 cm^{-1} towards lower frequencies is proportional to solvent polarity.

The authors determined C-H bond deformation oscillation shifts in chloroform in 25 solvents [43] and perform a correlation of shift values, affected by interaction with solvents, with the solvent properties. Under experimental conditions (10% solution in cyclohexane or pure preparation), the maximum of deformation oscillations at δ = 1210 cm^{-1} shifts towards higher frequencies that correlates with other results [44]. Moreover, in most cases the band configuration is changed compared with absorption in cyclohexane. Usually, the absorption intensity at the maximum decreases, by the band becomes wider. In some cases, a threshold is observed on the band (acetone, benzonitrile, dibutyl ester, dimethyl sulfoxide) or splitting to a doublet (triethylamine, dioxane) or triplet (diethylamine, quinoline, diethyl ether, piperidine). The appropriate shift values of C-H absorption band maximum ($\Delta\delta$, cm^{-1}) in studied solvents in relation to C-H absorption band maximum for 10% $CHCl_3$ in cyclohexane δ = 1210 cm^{-1} are shown in Table 3. Shift values are also shown for splitting of the most intensive bands.

However, generalization of data obtained from five-parameter equation gives none of any definite dependence. The equation is characterized by extremely low correlation (R = 0.737). Only elimination of the most deviating data for five solvents (butanol, aniline, piperidine, DMFA, and dioxane) may give an acceptable adequacy of the equation:

Table 3. Deformation oscillation C-H band shifts in chloroform (1210 cm^{-1}) under solvent influence

No.	Solvent	$\Delta\delta$	
		experimental	calculated
1.	*p*-Xylene	0	4.1
2.	Acetic anhydride	0	7.6
3.	Benzene	2	3.7
4.	Cyclohexanone	5	15.4
5.	Nitrobenzene	5	1.1
6.	Benzonitrile	5	8.8
7.	Acetonitrile	15	14.8
8.	Ethyl acetate	15	16.9
9.	Dibutyl ester	18	21.7
10.	Tetrahydrofuran	21	19.8
11.	Acetophenone	21	12.5
12.	Diethyl ester	24	23.2
13.	Dimethyl acetamide	25	31.7
14.	Dimethyl sulfoxide	32	25.8
15.	Dioxane	36	eliminate
16.	Dimethyl formamide	36	eliminate
17.	Dichloroethane	16	11.0
18.	Anisole	25	11.2
19.	*n*-Butanol	0	eliminate
20.	Aniline	4	eliminate
21.	Piperidine	23	eliminate
22.	Diethylamine	60	53.3
23.	Formamide	77	77.4
24.	Quinoline	30	33.2
25.	Triethylamine	50	52.8

$$\Delta\delta = 82.31 + (32.21 \pm 43.73)\frac{n^2-1}{n^2+2} - (69.17 \pm 37.88)\frac{\varepsilon-1}{2\varepsilon+1} -$$

$$- (0.081 \pm 0.022)\delta^2 + (0.081 \pm 0.009)B + (3.43 \pm 0.48)E_T; \qquad (36)$$

$R = 0.958$; $S = 6.84$.

After elimination of a low significant chargeability factor, we obtain:

$$\Delta\delta = -73.77 - (71.49 \pm 37.15)\frac{\varepsilon-1}{2\varepsilon+1} - (0.077 \pm 0.021)\delta^2 +$$

$$+ (0.08 \pm 0.09)B + (3.41 \pm 0.47)E_T; \qquad (37)$$

$R = 0.956$; $S = 6.7$.

The maximum paired coefficient of correlation ($r = 0.82$) is observed for polarity, but if this parameter is eliminated, adequacy of the three-parameter equation obtained

$$\Delta\delta = f(\delta^2, B, E_T)$$

is reduced slightly to $R = 0.945$. Of much greater importance for the interaction are specific solvation factors (elimination of basicity reduces R to 0.69, and electrophilicity – to 0.78). Thus chloroform molecule interacts with the solvent primarily forming a weak hydrogen bond with basic solvents: Cl_3C-H...B. Meanwhile, an electrophilic solvation of chloroform molecule by electron deficient carbon atom is possible. Both solvation types weaken C-H bond stretching it and, therefore, promote an increase of its spectral oscillation frequency, $\Delta\delta$. At the same time, self-association of solvent molecules impedes solvation processes, which is indicated by the sign "minus" at the appropriate term of the regression equation describing correlation between physicochemical parameters of the environment and $\Delta\delta$. A conclusion about determining effect of nucleophilic solvation of chloroform on its spectral characteristics is also confirmed by analysis of C-D valence band shifts ($\Delta\nu$) in $CDCl_3$ and chemical shifts in ESR spectra of chloroform.

Besides the initial experimental data on band shifts $\Delta\delta$ in $CHCl_3$, Table 3 shows values, calculated by equation (37). Clearly correlation between experimental and calculated values is good; deviations (as the mentioned solvents are excluded) are below 10 cm^{-1}.

Since the most of eliminated solvents represent proton-containing substances, it is not inconceivable that solvation processes are characterized by some features. Therefore, an attempt was made to generalize data on $\Delta\delta$ for proton-containing solvents and *tert*-amines (No. 19 – 25). For these solvents $R = 0.943$. After pyridine elimination the equation gives $R = 0.998$. Similar to the previous case, specific solvation factors are of high significance, but in this case relative contribution of electrophilic solvation is much higher rather than of nucleophilic one:

$$\Delta\delta = -546.10 + (0.48 \pm 0.04)B + (8.70 \pm 0.75)E_T; \tag{38}$$

$n = 6$; $R = 0.991$; $S = 5.4$.

Therefore, an acceptable correlation may be obtained only at separate consideration of unassociated and associated solvents. Besides compounds with hydrogen bonds, associated solvents include tertiary amines. It is proved by their self-association: for example, pyridine's boiling point is 30°C higher than for benzene, which is similar by structure and molecular weight. Unfortunately, even at such separation the dependence of $\Delta\delta$ on properties of unassociated solvents is characterized by rather low, though significant correlation:

$$\Delta\delta = -20.49 - 13.48\frac{n^2-1}{n^2+2} - 58.1\frac{\varepsilon-1}{2\varepsilon+1} +$$

$$+ 69.58\delta^2 + 0.0098B + 0.894E_T; \tag{39}$$

$R = 0.958$; $S = 6.84$;

$r_{01} = 0.294$; $r_{02} = 0.078$; $r_{03} = 0.459$; $r_{04} = 0.549$; $r_{05} = 0.258$.

It is also found that elimination of low significant parameters reduces R to 0.906.

The main effect on $\Delta\nu$ is caused by basicity B, i.e. nucleophilic solvation of chloroform by n- or π-donors. Other parameters are relatively insignificant, though generalization should take into account both nonspecific solvation induced by solvent polarity and purely physical interaction, characterized by parameter δ^2.

For unassociated solvents, positive regression indices at B and E_T indicate an increased shift towards higher frequencies with specific solvation. Primarily, specific solvation increases with solvent basicity, i.e. with the equilibrium shift towards complex formation. The effect of nonspecific solvation is opposite: surrounding solvent molecules may orient along C-Cl bonds. For associated solvents, the effects of both nonspecific and specific solvation are similar: they decrease $\Delta\nu$. Meanwhile, the observed shifts towards higher frequencies are caused by purely physical interactions, characterized by cohesive energy density.

These conclusions are confirmed by the data [45] on the solvent effects on valence oscillation band location at $\nu = 2553$ cm^{-1} for C-D bond in $CDCl_3$ (Table 4). The most of donor solvents induces a valence band shift towards lower frequencies.

Table 4. C-D bond valence oscillation shifts in $CDCl_3$ at 2553 cm^{-1} under the solvent effect [45]

No.	Solvent	Δν
1.	Diethyl ester	-10
2.	Dibutyl ester	-11
3.	Dioxane	-5
4.	Anisole	-3
5.	Acetone	0
6.	Acetic anhydride	+7
7.	Ethyl acetate	+3
8.	Pyridine	-36

Valence oscillation shifts of C-D bond induced by solvents may be adequately described by a two-parameter equation, taking into account only basicity and chargeability of the solvents:

$$\Delta\nu = 42.44 - 92.26\frac{n^2 - 1}{n^2 + 2} - 0.104B;\ R = 0.992;\ S = 2.01. \qquad (40)$$

Since C-D valence frequencies, contrary to deformation frequencies, decrease in interaction with the solvent, the regression index at basicity is negative.

The effect of nonspecific solvation of unionized proton, though it is capable of forming a hydrogen bond, on ESR chemical signal shift in chloroform is also clearly expressed [46]

(Table 5). The fullest data on chemical shifts for chloroform in 14 solvents [47] did not give an adequate correlation:

$$\delta_{H}^{1} = (0.229 \pm 0.080) + (0.0018 \pm 0.00028)B; \quad (41)$$

$R = 0.904$; $S = 0.186$.

Table 5. ESR chemical shifts for $CHCl_3$in solvents [47]

No.	Solvent	δ_{H}^{1}, ppm
1.	Cyclohexane	0
2.	CCl_4	0.128
3.	$CHCl_3$	0.178
4.	Dibutyl ester	0.538
5.	Dioxane	0.578
6.	Diethyl ester	0.680
7.	Ethyl acetate	0.714
8.	Diisopropyl ester	0.716
9.	Methyl acetate	0.723
10.	Tetrahydrofuran	0.765
11.	Cyclohexanone	0.914
12.	Methyl ethyl ketone	0.922
13.	Triethylamine	1.158
14.	Piperidine	1.608

Meanwhile, additional consideration of polarity and cohesive energy density parameters provides for good correlation:

$$\delta_{H}^{1} = -0.545 + 1.597\frac{\varepsilon - 1}{2\varepsilon + 1} + 0.0031 + 0.00182B; \quad (42)$$

$R = 0.986$; $S = 0.081$.

Therefore, though solvent basicity is the most important factor defining the value of ESR chemical shift in $CHCl_3$, satisfactory correlation between these values is obtained only at taking into account perturbing effects, induced by both self-association and polarity of the solvent. As cohesive energy density is excluded from the consideration reduces the coefficient of correlation to 0.979. Exclusion of polarity reduces it even lower – to 0.937. This testifies about correct one-parameter correlation for substances with weak hydrogen bonds.

Similar results were obtained in the study of the solvent effect on the shift of ESR signal from extremely weak proton donor – ethylcyanacetate [48]. Analysis of the environment effect on valence and deformation oscillations of C-H bond in IR-spectrum is complicated by the band overlapping. Meanwhile, NMR spectra give more satisfactory peak separation of C-

H bonds in ethylcyanacetate rather than IR-spectra. This allows the study of the solvent origin effect on chemical shift of protons in CH_2-group of ethylcyanacetate by the current method. Interpretation of NMR spectra and commensuration of observed peaks with ethylcyanacetate molecule structure allow relation of triplet with 1 ppm chemical shift to CH_3 group. The quartette located in weaker field is related to CH_2-group of the ester (chemical shift equals 3.2 ppm). Clearly separated single peak with 2.7 ppm chemical shift corresponds to hydrogen atoms of methyl group.

Table 6. The solvent origin effect on chemical shift of methylene group in ethylcyanacetate

Solvent	${}^{1}_{\Delta}H$, ppm
Proton solvents	
Cyclohexanol	4.0
Benzyl alcohol	2.7
Piperidine	2.2
Pyridine	3.7
Diethylamine	2.7
Quinoline	3.6
Aniline	2.8
Aprotic solvents	
Dimethyl sulfoxide	4.2
Triethyl phosphate	2.2
Dimethyl formamide	4.0
Acetonitrile	2.4
Ethyl acetate	2.6
Triethylamine	3.6
p-Xylene	2.9
1,4-Dioxane	2.1
Nitromethane	3.2
Ethylcyanacetate	2.7

Table 6 shows investigation results from [48]. However, the coefficient of correlation equals $R = 0.541$, which is unsatisfactory value. The solvent influence on the proton shift in active methylene group is well described by the correlation equation only at separate consideration of proton and aprotic solvents. For proton solvents the correlation between chemical shift ${}^{1}_{\Delta}H$ and characteristic parameters ($R = 0.990$) is described by the following multi-parameter equation:

$${}^{1}_{\Delta}H = 7.572 + 1.540\frac{\varepsilon-1}{2\varepsilon+1} - 3.292\frac{n^2-1}{n^2+2} - 0.267E - 0.0072B. \quad (43)$$

Since polarity and chargeability effect is insignificant, the influence of proton solvents on chemical shift of acidic methylene group in the ester is defined by specific solvation of ethylcyanacetate by the solvent only.

$$^{1}_{\Delta}H = 7.163 - 0.273\text{E} - 0.0072\text{B};\ \text{R} = 0.976. \qquad (44)$$

Hence, the increase of both acid and base solvation contribution from the solvent is accompanied by CH_2-group peak shift towards strong magnetic field indicating reduction of proton mobility in the methylene group. At electrophilic solvation by carbonyl oxygen of carboxylic group of cyanacetic ester, electrons are repulsed from methylene group that causes shrinkage of C-H bond.

Estimation of aprotic solvent effect on chemical shift in methylene group gave the following correlation dependence:

$$^{1}_{\Delta}H = -3.534 + 4.555\frac{\varepsilon - 1}{2\varepsilon + 1} + 17.314\frac{n^2 - 1}{n^2 + 2} + 0.0024B; \qquad (45)$$

$R = 0.944.$

This dependence takes into account the effect of both specific and nonspecific solvation of the ester by aprotic solvents. Electrophilicity is the insignificant parameter and may be excluded from the correlation above. Basicity characterizes specific solvation of active methylene group by nucleophilic solvent that induces an increase of active proton mobility and shift of CH_2 peak in NMR spectrum of the ester in a weaker magnetic field. As a consequence, electrophilic solvents do not cause any significant effect on ethylcyanacetate and transfer of electron effects, induced by interaction with donor C=O and C≡N groups of the ester, is insignificant for sensitivity to methylene group peak shift.

Thus total effect of the solvent on cyanacetate molecule is estimated by the dominating effect of specific solvation, displayed in formation of a hydrogen bond with carbonyl group and electron effect transmission to carbon atom of CH_2-group, as well as by interaction of active methylene group with solvent molecules.

Multiparameter equations are effective for generalizing data on solvent interaction with weak proton donors and explaining solvation processes for even much stronger donor – phenol. The use of multiparameter equation allows improvement of correlation between environment properties and ESR signal shift, induced by them [46].

As an example, the data on solvent effect on chemical shifts of proton signals in ESR spectrum of phenol. Points on the graph of $^{1}_{\Delta}H$ (*B*) dependence fit two parallel lines. To align these lines the authors suggest subtracting donor "anisotropy" increment from $^{1}_{\Delta}H$ for deviating points of strong donors (carbonyl compounds, pyridine). Meanwhile, data on 16 solvents possessing all parameters (Table 7), treated by a five-parameter equation, gave a satisfactory correlation ($R = 0.944$) with the paired coefficient of correlation with basicity $r = 0.909$. Elimination of data for the most deviating solvents (pyridine and acetonitrile) allows an increase of the coefficient of correlation to $R = 0.970$. Since the effect of electrophilicity is insignificant, the connection between phenol's $\delta^{1}_{\Delta}H$ and solvent parameters is properly described by the following four-parameter equation:

$$^{1}_{\Delta}H = 6.467 - 7.406\frac{n^2-1}{n^2+2} + 2.348\frac{\varepsilon-1}{2\varepsilon+1} + 7.813\delta^2 + 0.00953B; \quad (46)$$

$R = 0.969$; $S = 0.346$;

$r_{01} = 0.083$; $r_{02} = 0.043$; $r_{03} = 0.060$; $r_{04} = 0.910$.

Table 7. Chemical shift of phenol ESR signals [49]

No.	Solvent	$^{1}_{\Delta}H$, ppm
1.	Sulfolane	7.80
2.	Acetonitrile	7.30
3.	Ethylacetate	8.38
4.	Acetone	8.82
5.	Dioxane	8.32
6.	Trimethylphosphate	9.10
7.	Diethyl ether	8.52
8.	Tetrahydrofuran	8.83
9.	Dimethyl formamide	9.73
10.	N-methylpyrrolidone	10.25
11.	Triethyl phosphate	9.23
12.	Dimethyl acetamide	10.27
13.	Dimethyl sulfoxide	9.35
14.	Hexamethyl phosphamide	10.63
15.	Pyridine	11.93
16.	Triethylamine	11.88

A bit worse correlation is obtained by the three-parameter equation:

$$^{1}_{\Delta}H = 5.066 + 1.907\frac{\varepsilon-1}{2\varepsilon+1} + 6.418\delta^2 + 0.00914B; \quad (47)$$

$R = 0.958$; $S = 0.362$.

In some other cases, multiparameter equations allow quantitative generalization of results on NMR signal changes in various environments and their extended interpretation [46].

1.4. The Study of Iodine Complex UV-Spectra

Iodine is a classical object for studying charge transfer complexes (CTC), by UV-spectrometrical techniques, in particular. There are hundreds of works devoted to iodine complexes and their spectra in various solvents. Nevertheless, far not all aspects of the problem, mostly quantitative, are well determined. Therefore, the iodine-donor systems are of great interest for investigators still.

Variations in iodine solution colors with the solvent type are known since this element was discovered. Benesi and Hildebrand [50] studied this phenomenon quantitatively. They demonstrated that as a result of interaction with solvents and correspondingly, increase of donor electron mobility, iodine absorption band in the visible region of spectrum obtains a hypsochrome shift. At the same time, high intensity absorption band occurs in UV-region (at 300 μm), corresponded to absorption of formed CTC. As the donor strength increases, the maximum of this band is shifted towards longer wavelengths, i.e. opposite to the visible region. The investigations show [51] that also the long wave absorption band of iodine complexes, at least, in the case of *n*-donors, is associated with the complex formation and may be related to external complex (CTC-II) with the structure close to ionic type: $[DJ]^{+}J^{-}$. The structure and properties of iodine complex with donors are discussed in many monographs and reviews [53 – 56]. The question about structure and bond type in iodine complexes was the most ambiguous. Mallikan suggested that iodine molecule located parallel to aromatic donor plane [56]:

However, Keefer and Andrews [57] suggest more probable the structure, in which halogen molecule axis coincides with the hexad axis of the aromatic ring:

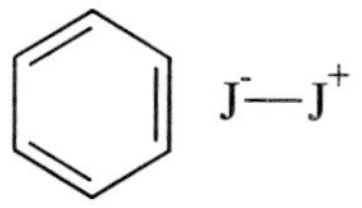

This structure was proved by IR-spectroscopic and crystal studies. Though this structure by Mallikan is less profitable, because it is characterized by calculated lower electron transfer, actually, other benzene orbitals take part in bond formation so that total overlap integral is high. The complexes are formed due to existence of low antibonding orbitals in iodine and their ability to polarize. Such coaxial structure was also detected for iodine complexes with olefins [58].

Of special attention is the question about the type of bond between iodine and donors, because, on the one hand, it is associated with CTC component properties, and on the other hand, defines characteristics of the complex. According to Mallikan, both charge transfer interactions and classical electrostatic forces of dipole-dipole, dipole-induced dipole interactions, etc. are displayed in CTC. It is shown in the known formula [59]:

$$\psi_N = a\psi_0(D, A) + b\psi_1(D^+ - A^-). \quad (48)$$

In later works the opinion dominated that consideration of donor-acceptor complexes necessitated taking into account various types of interactions: "…as a donor-acceptor bond is formed between two valence-saturated molecules, all forces are displayed to their higher extent, including Coulomb's, exchange, dispersion, as well as specific interaction induced by electron transfer or displacement from donor to acceptor. Each factor makes separate contribution into total effect of molecular compound formation" [54, p. 27].

For iodine complexes, primarily it was found that linear dependencies between visible (or UV) absorption band shift and donor ionizing potential existed with series of similar compounds: for aromatic hydrocarbons and haloalkanes [52, pp. 29, 33], phenyl ethers, thioethers and selenoethers [60], and aromatic hydrocarbons and alkenes [61].

Similarly, other properties of iodine complexes, such as calorimetrically determined heats of mixing [62] or iodine solubility [63] show linear correlation with ionization potential or induction factor in the Hammett equation only within short series of similar compounds.

Of special interest is a correlation suggested between parameters of donor complex formation with iodine or other acceptors. Gutmann [3, p. 28] reports on linear correlation between spectrally determined enthalpies of complex formation by some donors with phenol or iodine and donor numbers *DN* (i.e. enthalpies of mixing with $SbCl_5$). However, Drago [4, p. 105] showed that such conclusion was incorrect and made on the basis of insufficient data – there is no linearity but some symbate property between these values. Spectrally determined free energies of complex formation with iodine give three linear dependencies on the donor basicity: for oxygen-, selenium- and sulfur-containing donors, separately [64]. Exceptions from linear dependencies of lg$k$ on the donor basicity are also indicated [65, 66].

It is shown [51] that for n-donors, long-wave absorption band shift in iodine complex (CTC-II) spectrum is linear with respect to donor ionizing potential and their Palm basicity B. Meanwhile, no similar correlation was obtained for short-wave band shift for iodine-n-, π-donor systems.

The absence of strict linear correlation between energy parameters of iodine complex formation and complexes with hydrogen bonds (including phenol) is indirectly confirmed [4, 62] by suggesting enthalpy of donor complex formation with iodine as one of parameters of multiplicative equation that characterizes enthalpies of complex formation with other acceptors. To characterize equilibrium constants of complex formation with hydrogen bonds [68], a two-parameter equation is suggested:

$$\lg k_{pi} = A + B_p k_{HB} + C \lg k_{J_2}, \quad (49)$$

where the first term characterizes proton transfer and the second term – charge transfer.

To avoid such nonconformities, specific and nonspecific interactions should be taken into account in the system, simultaneously. Moreover, their share participation may be different with respect to the type of components. Such point of view was supported in the work [69], where interaction parameters of iodine-amines complex formation represent a sum of terms, both associated and unassociated with the charge transfer. Other authors [70] accept universal interactions to be of the highest importance: "even in spectra affined to specific interactions of molecules, to which iodine molecule also relates, universal intermolecular forces are displayed quite typically, clearly affecting the entire image of intermolecular interaction. In such cases, determination of the above-mentioned universal dependence of high importance for stipulated consideration of specific interaction represents an important stage of analysis of the experimental material".

1.4.1. Study of the Solvent Effect on Iodine CTC Parameters

The dependence of constants and enthalpies of the complex formation on the type of environment is known for a long time [54]. For iodine complexes Yu.Ya. Fialkov showed that k and ΔH values of iodine interaction with dimethyl acetamide, pyridine, benzene, and dimethyl ether were quite different even in hydrocarbon solvents [71] (Table 8).

Table 8. Dimethyl acetamide-iodine complex parameters in various solvents [71]

Solvent	k_{25}, l/mol	ΔH	
		kcal/mol	kJ/mol
Cyclohexane	16.5 ± 6.2	19.8 ± 1.1	82.76 ± 4.598
n-Hexane	14.3 ± 0.2	19.8 ± 1.5	82.76 ± 6.27
Benzene	2.58 ± 0.11	56.85 ± 0.5	56.85 ± 2.09
m-Xylene	1.31	10.3	43.054
p-Xylene	1.34	10.7	44.726

Table 9. Constants and enthalpies of pyridine-iodine complex formation in various solvents

References	k, l/mol				$-\Delta H$, kcal/mol			
Solvent	[72]	[73]	[75]	[81]	[72]	[73]	[75]	[81]
n-Hexane	147	148	-	-	8.5	7.3	-	-
CCl_4	111	102	105	-	7.0	8.6	7.5	-
Benzene	91	-	82.9	76.5	7.1	-	-	6.9
Toluene	-	-	87.8	-	7.6	-	-	-
Chlorobenzene	196	-	-	-	8.6	-	-	-
Bromobenzene	221	-	-	-	5.8	-	-	-
Chloroform	-	-	77.7	-	-	-	7.8	-
Dichloromethane	-	-	151	-	-	-	8.6	-
Dichloroethane	-	-	189	-	-	-	7.8	-
Cyclohexane	-	135	-	-	-	8.6	-	-
n-Heptane	-	142	157	147	-	7.4	8.2	8.12
n-Xylene	-	91.6	62.6	-	-	-	-	4.28

Similar results were obtained for iodine complexes with pyridine [72 – 75] and 3-chloropyridine [76] by spectrophotometry technique and confirmed calorimetrically [77, 78]. The reason for such dependence of constant values on the environment may be competitive formation of weak complexes between iodine and the solvent: it is shown [70] that iodine spectrum is different even for solutions in hexane, heptane and cyclohexane. As a consequence, complex formation constants also differ: 0.15 for benzene, 0.0127 for cyclohexane, and 0.0034 l/mol for 2,3-dimethylbutane. Iodine complex formation with alkanes, CCl_4, $CHCl_3$ and dichloroethane is also proved by the solubility method [60], though obtained values of constants are higher. The opinion seems to be more valid [71, 78, 81] that the equilibrium state is affected by solvation of all participants of the process - iodine, donor and a complex.

Generalization of data on Pyr-J_2 complex (Table 9) [71 – 75], approximately linear dependencies was only determined between k (or lg$k$) and dielectric constant ε or cohesive energy density δ^2 of the solvents.

As 11 values of lg$k$ were processed by a five-parameter linear equation, an expression with the coefficient of correlation $R = 0.938$ was obtained. Toluene was excluded as the most highly deviating component (No. 4 [75]), and correlation was increased to a satisfactory level:

$$\lg k = 2.948 - 5.586f(n) + 1.164f(\varepsilon) + 0.0047\delta^2 + 0.0067B - 0.154E; \quad (50)$$

$R = 0.975$; $S = 0.050$;

$r_{01} = 0.292$; $r_{02} = 0.454$; $r_{03} = 0.301$; $r_{04} = 0.364$; $r_{05} = 0.298$

and

$$\lg k = 2.938 - 4.690f(n) + 1.580f(\varepsilon) + 0.0070B - 0.137E; \quad (51)$$

$R = 0.966$; $S = 0.052$.

Here the most valuable parameters are polarity and electrophilicity of the solvents. As other parameters are excluded from this four-parameter equation, the correlation is perturbed. The complex becomes more stable with increase of solvent polarity and basicity, because charge separation in the complex and its solvation are promoted. It should be taken into account that the study was performed only in low-donor solvents with $B < 60$. For solvents with higher donor property, one may expect k decrease as a result of competitive interaction between solvents and iodine. Chargeability and specifically electrophilicity reduce the complex stability due to electrophilic solvation of pyridine molecules by solvent. For example, the minimum k value is observed for the most electronegative chloroform. It is known that chloroform forms complexes with some organic donors, including pyridine [82, 83].

Worse correlation is obtained for the complex formation enthalpy ΔH: for 11 points R = 0.682. However, as data on cyclohexane, bromobenzene and n-heptane are eliminated subsequently, R is increased to 0.802, 0.934 and 0.990, respectively. The final data for 8 solvents may be described by the following equation:

$$-\Delta H = 19.47 - 7.943\frac{n^2-1}{n^2+2} + 12.83\frac{\varepsilon-1}{2\varepsilon+1} - 0.074\delta^2 + 0.018B - 0.081E; \quad (52)$$

$R = 0.990$; $S = 0.17$;

$r_{01} = 0.339$; $r_{02} = 0.427$; $r_{03} = 0.059$; $r_{04} = 0.139$; $r_{05} = 0.082$.

The exclusion of insignificant parameters (chargeability and electrophilicity) does not deteriorate the correlation, and the environment effect on enthalpy of pyridine-iodine complex formation may be described by the three-parameter equation:

$$-\Delta H = 10.45 + 13.65\frac{\varepsilon - 1}{2\varepsilon + 1} - 0.085\delta^2 + 0.0134B; \tag{53}$$

$R = 0.976$; $S = 0.183$.

Therefore, the solvent effects on pyridine-iodine complex formation are satisfactorily described by multiparameter equations of linear free energies, the most valuable being the factors of nonspecific solvation and cohesive energy density. Sometimes, it may partly be stipulated by features of the studied system. Only solvents with low or zero basicity are studied, whereas solvents with higher donor property (ethers, for example) will necessarily solvate the pyridine-iodine complex, and the competition for acceptor (iodine) molecule between donor solvent (D_{solv}) and pyridine will occur. This will completely disturb the picture of UV absorption spectrum:

$$D_{solv} + Pyr{\cdot}J_2 \leftrightarrow D_{solv}{\cdot}J_2 + Pyr. \tag{54}$$

This was the reason for studying [71 – 79] only few low-basicity solvents, which reduced significance of the generalization performed. However, the effect of about 30 solvents on $\pi\to\sigma^x$ and $\sigma\to\sigma^x$ transition energies was studied [84] for anions J_3^- in CsJ_3, $[Ph_3As]J_3$ and $[Et_4N]J_3$ (Table 10). The authors observed a satisfactory linear dependence between these energies. However, they cannot be correlated by one-parameter dependencies with parameters DN, δ^2 or $\frac{n^2 - 1}{n^2 + 2}$.

Primarily, data on the transition $\pi\to\sigma^x$ in CsJ_3 (Table 10) processed by five-parameter equation gave poor correlation $R = 0.820$. Successive exclusion of ethylene glycol, methyl cellosolve, *tert*-butanol and isopropanol increases the coefficient of correlation to 0.879, 0.929, 0.955, and 0.963, respectively. The equation for 22 points is:

$$E_1 = 341 - 21.6\frac{n^2 - 1}{n^2 + 2} + 6.35\frac{\varepsilon - 1}{2\varepsilon + 1} + 0.003\delta^2 - 0.0017B + 0.33E; \tag{55}$$

$R = 0.963$; $S = 0.268$;

$r_{01} = 0.511$; $r_{02} = 0.157$; $r_{03} = 0.719$; $r_{04} = 0.514$; $r_{05} = 0.872$.

Table 10. Transition $\pi \rightarrow \sigma^x$ (E_1) and $\sigma \rightarrow \sigma^x$ (E_2) energies for J_3^- ion in CsJ_3 [84]

No.	Solvent	E_1, kcal/mol	E_2, kcal/mol
1.	Water	81.5	99.4
2.	MeOH	79.6	98.2
3.	EtOH	79.4	98.0
4.	*n*-PrOH	79.3	98.0
5.	*n*-BuOH	79.4	97.7
6.	*iso*-PrOH	79.5	98.0
7.	*tert*- BuOH	79.6	98.1
8.	$PhCH_2OH$	78.6	-
9.	$OHCH_2CH_2OH$	78.2	96.8
10.	$MeOCH_2CH_2OH$	77.8	96.7
11.	$CHCl_3$	78.7	97.5
12.	CH_2Cl_2	78.9	97.4
13.	$HCONH_2$	78.1	96.1
14.	Me_2SO	78.0	96.2
15.	MeCOOEt	78.7	97.7
16.	$MeCONMe_2$	78.0	96.4
17.	Me_2CO	78.4	-
18.	$(MeCO)_2O$	78.7	97.5
19.	$(BuO)_2PO$	77.9	96.8
20.	$(Me_2N)_3PO$	77.6	96.2
21.	MeCN	79.0	98.0
22.	Pyridine	77.6	-
23.	Dioxane	78.8	-
24.	Tetrahydrofuran	77.9	96.7
25.	Et_2O	78.4	-
26.	$MeOCH_2CH_2OMe$	78.0	96.8

Among all parameters, polarity is of the lowest significance:

$$E_1 = 331 - 21.9\frac{n^2 - 1}{n^2 + 2} + 0.005\delta^2 - 0.00019B + 0.32E; \qquad (56)$$

$R = 0.958$; $S = 0.277$.

Further elimination of parameters will just reduce R to 0.90 – 0.93.

Similar to E_1, after exclusion the same points for E_2, the rest 18 solvents give good correlation by 4- and 3-parameter equations:

$$E_2 = 99.03 - 6.03f(n) + 0.001\delta^2 - 0.00255B + 0.073E; \qquad (57)$$

$R = 0.962$; $S = 0.27$;

$r_{01} = 0.708$; $r_{03} = 0.555$; $r_{04} = 0.573$; $r_{05} = 0.829$.

$$E_2 = 98.91 - 6.19f(n) - 0.0024B + 0.089E; \quad (58)$$

$R = 0{,}958$; $S = 0.27$.

Thus, for complex anion J_3^-, solvent effects on the complex stability are satisfactorily described by linear free energy, multiparameter equations. Signs at regression terms are the same as in equation for pyridine-iodine complex formation enthalpy. Spectral transition energy increases with cohesive energy density and electrophilicity of the solvent and decreases with polarity and basicity [84].

1.4.2. Application of LFE Equations to Setting a Correlation Between Donor Properties and Parameters of Donor-Iodine Complexes

The paramount material present in the literature on iodine CTC is generalized in the monograph [54]. Unfortunately, there are no quantitative dependencies allowing correlation of iodine CTC $\Delta\nu$, lg$k$, and ΔH values with donor properties, useful for all types of donors. With respect to the above-said, such correlation is possible only with taking into account all types of interactions in the system. Therefore, a question is brought up, if LFE equations developed for solvation effects of the medium are valid for considering the effect of properties of one of the components. Two points should be discussed. Firstly, the LFE principle is useful for taking into account the features of reagent structure in kinetic and spectral processes. Secondly, the most important is that spectrophotometry determination of constants and enthalpies of the complex formation using Benesi-Hildebrand, Rose-Drago, Ketelaare, etc. techniques was performed in the presence of highly excessive donors. The same can be said about determination of spectral shifts. It is logical to suggest that selective solvation of acceptor by excessive donor, stronger than the solvent, will dominate in the system. Usually, the solvent is rather inert (alkane, CCl_4), but determination of absolute k values does not allow neglecting of even low-active environment. Such suggestion was made [85] to explain changes of the complex formation constant with concentration for trinitromethane and aromatic hydrocarbons, determined by NMR technique. Later on, formation of solvation shells around the complex was also detected in the case of complexes with hydrogen bonds [86, 87]. The authors of these works conclude about an existence of a definite concentration zone, where spectrophotometry techniques applied to determination of equilibrium constants give reliable values; outside this zone, the values of constants begin increasing with dilution, as a result of the "cell effect".

To check if such selective solvation by excessive donor is displayed for iodine complexes, dimethyl formamide-iodine complex was studied by UV-spectrophotometry technique in wide range of concentrations [89].

Iodine in cyclohexane possesses by two absorption bands: one intensive, narrow and asymmetric in UV-region of the spectrum with the maximum at 230 – 235 nm (in the range of 145 – 240 nm [55]), and another low-intensive and wide, in the visible spectrum with the maximum at 520 nm [89]. The UV-region (220 – 230 nm) also displays a narrow, intensive absorption band for DMFA, also typical of other amides ($n \rightarrow \pi^x$ transition).

For iodine solution in DMFA (pure DMFA for test) a noticeable decrease of band intensity at 235 nm is observed. It is smeared and shifted towards longer waves: for 0.001M iodine in DMFA, optical density in the maximum at 295 nm is four-fold lower than for

0.0005M iodine in cyclohexane. Similarly, the visible absorption band of iodine loses its intensity, but shifts towards shorter waves – the absorption maximum is observed at 370 nm.

Figure 1 shows a gradual shift and iodine absorption band attenuation in the visible region (520 nm, 0.001 M solution in cyclohexane) with DMFA concentration increase from 0.01 to 0.1 mol/l (i.e. from 10- to 100-fold excess in relation to iodine).

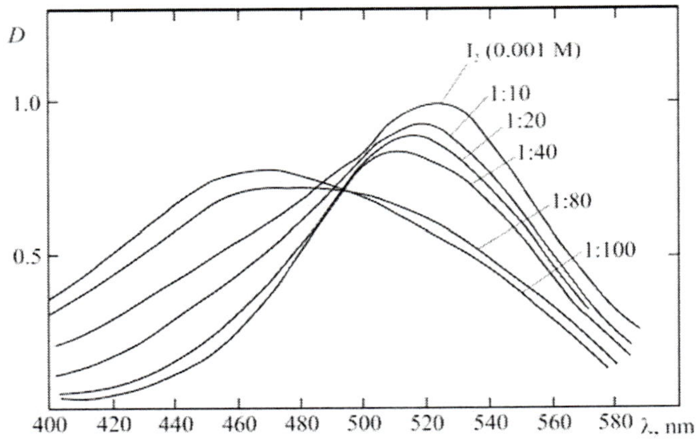

Figure 1. Absorption band shift of iodine in cyclohexane ($\lambda = 520$ nm) in the presence of various quantities of dimethyl formamide [88]

The equilibrium constant of complex formation, *k*, was determined using waves with $\lambda =$ 400 – 460 nm, where the complex is the main absorber, and free iodine absorption is negligibly low. It is found that optical density *D* fits the Benesi-Hildebrand dependence:

$$\frac{a \cdot l}{D} = \frac{1}{\varepsilon_{AB} k} \cdot \frac{1}{b} + \frac{1}{\varepsilon_{AB}}, \quad (59)$$

i.e. linearity in $\frac{C_{J_2} \cdot l}{D} - \frac{1}{C_{DMFA}}$ coordinates is observed (Figure 2). Moreover, calculated *k* value decreases with λ almost by an order of magnitude at constant extinction coefficient (Table 11).

Table 11. Complex formation constants k and extinction constants E for DMFA-iodine system, determined from the Benesi-Hildebrand equation for different wavelengths

λ	k, l/mol	ε
400	7.2	714
410	8	1,000
420	11.3	1,000
430	15.6	1,000
450	35.7	1,000
460	50	1,000

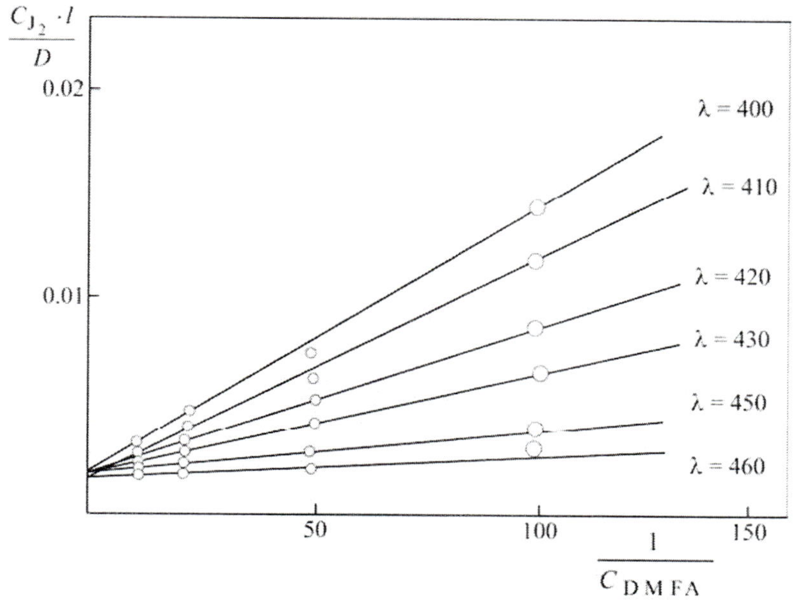

Figure 2. The application of Benesi-Hildebrandt equation to DMFA – iodine system for various wavelengths [88]

Calculation of constants by the Scott equation indicated the same changes. Such variability of k with wavelength was reported by many authors [90].

Such dependence of the complex formation constant k, as well as ΔG and ΔH, on the wavelength may be one of the reasons for reduced correlation between ΔG (or ΔH) and solvent parameters and deviation in ΔH values, reported for the same system by different investigators [54]. However, there may be another reason for this phenomenon. It is indicated that k values, determined for diluted solutions, may depend on the component concentration due to changes in activity [91] or formation of microaggregates in the solution [85].

Yu.L. Haldna et al. [92] suggested mathematical method for processing experimental UV-spectrophotometry data which allow separate consideration of donor-acceptor absorption component and, therefore, elimination of constant dependence on the wavelength. The value obtained (k = 7.78 l/mol) conform well to the range determined for 400 – 410 nm: k = 7.2 – 8 l/mol.

Table 12. Complex formation constants in DMFA-iodine system at different iodine concentrations

C_{J_2}, mol/l	lg*C*	*k*, l/mol	lg*k*
0.01	-2.0	250	2.40
0.005	-2.7	750	2.87
0.001	-3.0	3,750	3.57
0.0005	-3.7	5,100	3.71
0.0001	-4.0	6,125	3.79

In the DMFA-iodine system, k sharply increases as iodine concentration decreases (Table 12). These k values were calculated from the change of integral absorption density in the visible region at iodine:DMFA = 1:1 using the known technique [93].

The dependence of lgk on lgC is approximately linear, which is also observed for systems with halohydrocarbons [94].

Data on enthalpies of complex formation with iodine (ΔH_{J_2}) were taken from E.N. Gurianov *et al.* monograph [54, pp. 190, 315] for generalization. Unfortunately, among 150 substances, shown in this book, only data for 27 compounds may be used - those for which all required parameters are present in the literature. If there are several values of ΔH, newer UV-spectroscopy data are preferable. Data shown in Table 13 are taken from the most representative work [54].

There was deduced a five-parameter linear equation [95] binding complex formation enthalpy with the solvent properties, giving the coefficient of correlation R = 0.953. Elimination of mostly deviating point (No. 27, ethylamine) increased R to 0.968.

$$-\Delta H = 0.608 + 7.816\frac{n^2-1}{n^2+2} - 1.006\frac{\varepsilon-1}{2\varepsilon+1} -$$

$$- 0.019\delta^2 + 0.0143B - 0.076E; \quad (60)$$

$R = 0.968$; $S = 0.717$;

$r_{01} = 0.123$; $r_{02} = 0.048$; $r_{03} = 0.142$; $r_{04} = 0.907$; $r_{05} = 0.138$.

Low values of r_{0i} and appropriate regression coefficients indicate low significance of polarity and electrophilicity. Finally, the effect of the solvent origin on ΔH is satisfactorily described by three-parameter linear equation:

$$-\Delta H = 1.037 + 4.782\frac{n^2-1}{n^2+2} - 0.016\delta^2 + 0.0140B; \quad (61)$$

$R = 0.965$; $S = 0.731$.

Further elimination of parameters leads to more noticeable decrease of total coefficient of correlation: $R = 0.962$ for $f(\delta^2, B)$; $R = 0.912$ for $f(n, B)$; $R = 0.197$ (!) for $f(n, \delta^2)$. Therefore, the main effect on enthalpies of complex formation from iodine and solvents (ΔH_{J_2}) is caused by the acid-base interaction of iodine with the donor solvent. The correlation degree is also affected by cohesive energy density (δ^2) and, to a lesser extent, solvent chargeability. Negativity of regression coefficient at δ^2 shows that the higher cohesive energy density of the donor is, the lower (all other factors being the same) ΔH_{J_2} will be. Obviously, formation of a complex requires injection of iodine molecule inside the donor microaggregate structure, linked by intermolecular forces, in the solvent and overcoming donor molecule association forces (in spectrophotometry techniques determining equilibrium constant the donor concentration is ten- or even hundred-fold exceeds the acceptor concentration). This interpretation conforms to the lattice model of liquid structure by Ya.I. Frenkel [96] and the assumption of microheterogeneous structure of solution [85].

Table 13. Values ΔH_{J_2} and ΔG_{J_2} for iodine complex formation with the solvents

No.	Solvent	$-\Delta H$		$-\Delta G$	
		kcal/mol	kJ/mol	kcal/mol	kJ/mol
1.	Diethyl ether	4.3	18.0	1.31	5.48
2.	Tetrahydrofuran	5.3.	22.2	1.70	7.11
3.	$MeOCH_2CH_2OMe$	3.7	15.5	0.10	0.41
4.	Dioxane	3.8	15.9	-0.05	-0.20
5.	Methanol	1.9	7.9	0.91	3.80
6.	Ethanol	4.5	19.2	-0.09	-0.37
7.	*tert*-Butanol	3.4	14.2	1.43	5.98
8.	Ethyl acetate	2.5	10.5	-	-
9.	Ammonia	4.8	20.1	2.4	10.3
10.	Triethylamine	12.0	50.2	5.0	20.9
11.	Diethylamine	9.7	40.6	4.2	17.56
12.	Pyridine	7.5	31.4	3.03	12.66
13.	Quinoline	7.2	30.1	2.82	11.79
14.	Acetonitrile	1.9	7.9	-0.59	2.38
15.	Dimethyl sulfoxide	3.6	15.05	1.43	5.98
16.	Dimethyl formamide	3.7	15.5	0.63	2.63
17.	Dimethyl acetamide	3.9	16.3	1.14	4.76
18.	Benzene	1.45	6.1	0.3	1.25
19.	Toluene	1.8	7.53	0.48	2.00
20.	*o*-Xylene	2.0	8.36	0.61	2.55
21.	*p*-Xylene	2.18	9.11	0.69	2.88
22.	Mesitylene	2.86	11.95	-	-
23	Fluorobenzene	1.4	5.85	-	-
24.	Chlorobenzene	1.19	4.97	-0.24	-4.19
25.	Bromobenzene	1.88	7.85	-	-
26.	Iodobenzene	3.2	13.37	-	-
27.	Ethylamine	7.4	30.93	3.7	15.47
28.	Trimethylene oxide	6.4	26.75	1.93	8.06
29.	Tetrahydropyran	4.9	20.48	1.66	6.93
30.	Phenetole	0.5	2.08	-	-
31.	Methyl acetate	2.5	10.45	-	-
32.	Dimethyl aniline	1.83	7.66	1.88	7.52
33.	Sulfolan	2.2	9.19	-	-
34.	Tetramethylurea	4.4	18.39	-	-
35.	Triethyl phosphate	3.2	13.37	1.1	4.59
36.	Tributyl phosphate	2.9	12.12	1.8	7.52

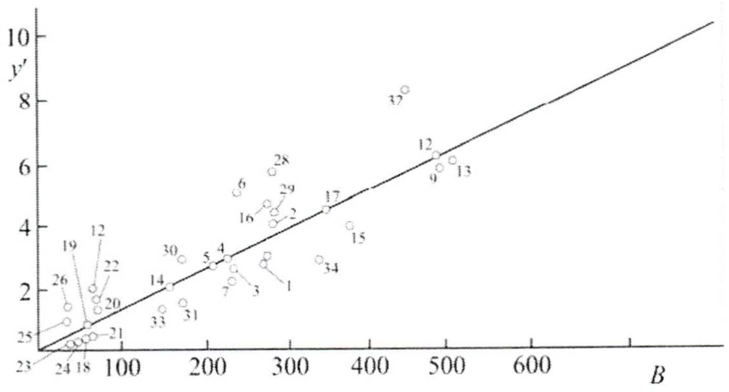

Figure 3. The effect of donor properties on enthalpy of complex formation with iodine. Dependence of y' on solvent basicity B. Points correspond to Table 13 [95]

Figure 3 shows dependence

$$y' = -\Delta H - 4.78f(n) + 0.016\delta^2 = f(n).$$

Noticeable deviations are observed only for tetrahydrofuran (No. 28) and dimethylaniline (No. 32).

Similar data processing on ΔG (points No. 1 – 7, 9 – 17, 27 – totally, 17 points) leads to an equation with R = 0.954. After elimination of data on ethylamine (No. 27) R increases to 0.968:

$$-\Delta G_{J_2} = -2.436 + 1.255\frac{n^2-1}{n^2+2} + 2.777\frac{\varepsilon-1}{2\varepsilon+1} -$$

$$-\ 0.0085\delta^2 + 0.00996B + 0.046E; \qquad (62)$$

$R = 0.968$; $S = 0.482$;

$r_{01} = 0.221$; $r_{02} = 0.391$; $r_{03} = 0.405$; $r_{04} = 0.510$; $r_{05} = 0.386$.

Chargeability and electrophilicity are found insignificant. After their elimination the following three-parameter equation is deduced:

$$-\Delta G_{J_2} = -2.040 + 2.360\frac{\varepsilon-1}{2\varepsilon+1} - 0.0058\delta^2 + 0.00975B; \qquad (63)$$

$R = 0.964$; $S = 0.463$.

For two-parameter equations R values will be the following: R = 0.961 for $f(\delta^2, B)$; R = 0.952 for $f(\varepsilon, B)$; R = 0.429 for $f(\varepsilon, \delta^2)$.

Therefore, similar to ΔH_{J_2}, donor basicity and cohesive energy (the factor preventing the complex formation) are determining parameters for ΔG_{J_2}. Clearly the value $\Delta G = \Delta H - T\Delta S$ with eliminated entropic term, significance of parameter δ^2 is much lower, whereas B is much more significant than for ΔH: paired coefficients of correlation r_{OB} equal 0.405 and 0.951, respectively.

The calculation results allow a sensible explanation why data on sulfur and selenium compounds deviate from linear dependence of ΔH_{J_2} [4] or ΔG_{J_2} [64] on $\Delta \nu_{PhOH}$ (donor basicity) or on one another [97]: for these compounds δ^2 are significantly different from values for their oxygen analogs and, correspondingly, their corrections will be higher.

As the absence of correlation between ΔH_{J_2} and ΔH_{PhOH} is reported [4], applicability of linear free energy equation to ΔH_{PhOH} [3] was tested. The following equation was deduced:

$$-\Delta H_{PhOH} = 5.791 - 4.579\frac{n^2-1}{n^2+2} - 0.013\delta^2 + 0.0171B + 0.0878E; \qquad (64)$$

$R = 0.970$; $S = 0.674$;

$r_{01} = 0.583$; $r_{02} = 0.119$; $r_{03} = 0.450$; $r_{04} = 0.943$; $r_{05} = 0.558$.

Chargeability and electrophilicity are relatively insignificant, therefore:

$$-\Delta H_{PhOH} = 4.554 - 11.386\frac{\varepsilon-1}{2\varepsilon+1} + 0.011\delta^2 + 0.0149B; \qquad (65)$$

$R = 0.968$; $S = 0.568$.

Thus both ΔH_{J_2} and ΔH_{PhOH} are satisfactorily correlated with the donor properties; the determining parameters being basicity and, to a lesser extent, cohesive energy density. However, regression coefficient values at separate parameters are different that makes one-parameter linear dependence between these values impossible.

As enthalpies and equilibrium constants, calculated from spectroscopic data, may be significantly different with respect to experimental conditions: wavelength of measurement, solvent, concentration and calculation method – it is desirable to check applicability of the LFE principle to processing of simpler values, which are iodine absorption band shifts in UV-spectrum.

Calculations were performed for visible region [51, 89] and E.I. Gurianov *et al.* [54, pp. 204 – 210] UV band. Corresponded λ and ν values are shown in Tables 14 and 15. Calculations were performed not for spectral frequencies themselves, but for absorption frequency shifts $\Delta\nu$ in relation to frequency in a "neutral" solvent, *n*-hexane: long-wave λ = 523 nm (ν = 1,912·10^3 m^{-1}) and short-wave λ = 238 nm (ν = 4,202·10^3 m^{-1}). Note also that contrary to data on the visible region, the short-wave spectrum was usually measured for iodine solutions in heptane, cyclohexane or CCl_4, only in the presence of definite amounts of donor that necessarily affects $\Delta\nu$.

Table 14. Iodine absorption band shifts in the visible region in different solvents

No.	Solvent	λ, nm	$\nu \cdot 10^3$, m^{-1}	$\Delta\nu$
1.	*n*-Hexane	523	1912	0
2.	CCl_4	517	1934	22
3.	Chloroform	512	1953	41
4.	Methylene chloride	506	1976	64
5.	Dichloroethane	497	2012	100
6.	Ethyl chloride	494	2024	112
7.	*tert*-Butyl chloride	495	2020	108
8.	Ethyl iodide	478	2092	180
9.	Benzene	502	1992	80
10.	Chlorobenzene	508	1969	57
11.	Bromobenzene	503	1988	76
12.	Toluene	496	2016	104
13.	Iodobenzene	493	2028	116
14.	Mesitylene	189	2045	133
15.	Anisole	490	2041	129
16.	Methanol	440	2273	361
17.	Ethanol	443	2257	345
18.	*n*-Propanol	445	2247	335
19.	*iso*-Propanol	446	2242	330
20.	Dimethoxyglycol	458	2183	271
21.	Diethyl ether	466	2146	234
22.	Di-*n*-propyl ester	470	2128	216
23.	Diisopropyl ester	472	2129	207
24.	Dioxane	452	2212	300
25.	Tetrahydrofuran	446	2242	330
26.	Ammonia	430	2326	414
27.	Pyridine	422	2370	458
28.	Diethylamine	410	2439	527
29.	Triethylamine	414	2415	503
30.	Dimethyl acetamide	440	2273	351
31.	Fluorobenzene	407	1973	61
32.	Ethyl acetate		2190	278
33.	Acetone		2200	288
34.	Water		2220	308
35.	*n*-Butanol	448	2230	318
36.	Dimethyl formamide		2240	328
37.	Piperidine		2420	508

Table 15. Iodine absorption band shift in UV region in different solvents in relation to absorption in h-heptane (λ = 238 nm) [54]

No.	Solvent	λ, nm	$\nu \cdot 10^3$, m^{-1}	$\Delta\nu$
1.	Methanol	231	4329	127.3
2.	Ethanol	233	4292	90.2
3.	*n*-Propanol	236	4237	35.6
4.	*i*-Propanol	236	4237	35
5.	*n*-Butanol	237	4219	17
6.	*tert*-Butanol	233	4292	90
7.	Diethyl ether	250	4000	-202
8.	Diisopropyl ether	255	3922	-280
9.	Tetrahydrofuran	249	4016	-186
10.	Dioxane	264	3788	-414
11.	Acetone	242	4132	-69
12.	Cyclohexanone	253	3953	-249
13.	Ammonia	229	4367	165
14.	Ethylamine	246	4065	-137
15.	Diethylamine	260	3846	-356
16.	Triethylamine	278	3597	-605
17.	Piperidine	260	3846	-356
18.	Pyridine	235	4255	53
19.	Quinoline	238	4202	0
20.	Dimethyl sulfoxide	276	3623	-579
21.	Benzene	292	3425	-777
22.	*p*-Xylene	303	3300	-901
23.	Mesitylene	330	3030	-1172
24.	Tolyene [52, p. 40]	302	3311	-890
25.	*o*-Xylene [52, p. 40]	316	3165	-1037
26.	*m*-Xylene [52, p. 40]	318	3145	-1057
27.	Cyclohexane [55]	302	3311	-890
28.	Acetone [55]	255	3922	-280
29.	Water [55]	190	5263	1061

Calculation of visible frequency shifts for all 37 points leads to an equation with R = 0.969. Elimination of maximum deviating point No. 32 (ethylacetate) increases R to 0.974. The following equation was obtained:

$$\Delta\nu \cdot 10^3 = -74.56 + 221.4\frac{n^2-1}{n^2+2} + 223.6\frac{\varepsilon-1}{2\varepsilon+1} -$$

$$- 0.114\delta^2 + 0.725B + 8.608E; \qquad (66)$$

$R = 0.974$; $S = 37.24$;

$r_{01} = 0.583$; $r_{02} = 0.119$; $r_{03} = 0.450$; $r_{04} = 0.943$; $r_{05} = 0.558$.

Thus, data on the determining effect of solvent basicity on spectral shift in the visible region for iodine [51] were proved on numerous solvents. Current paired coefficient of correlation ($r_{04} = 0.916$) is lower than that in [51] ($r = 0.970$), because we take more solvents, including alcohols affined to self-association. In this case, chargeability and cohesive energy are relatively insignificant parameters (the opposite effect, as in the case of ΔH_{J_2} and ΔG_{J_2}) and have no effect on $\Delta\nu$, which is described well by a three-parameter equation. Two-parameter equation gives slightly worse correlation:

$$\Delta\nu\cdot 10^3 = -16.12 + 210.6\frac{\varepsilon - 1}{2\varepsilon + 1} + 0.714B + 6.28E; \quad (67)$$

$R = 0.972$; $S = 37.00$;

$$\Delta\nu\cdot 10^3 = -54.27 + 0.726B + 8.12E; \quad (68)$$

$R = 0.966$; $S = 40.60$;

$$\Delta\nu\cdot 10^3 = -51.68 + 0.711B + 397.7\frac{\varepsilon - 1}{2\varepsilon + 1}; \quad (69)$$

$R = 0.954$; $S = 47.3$.

It is noteworthy that signs at all significant regression terms are positive: both nucleophilic and electrophilic solvation of iodine complexes confirm the opinion [51] about partly ionized form of CTC-II. Solvent polarity increase will promote charge separation in the complex.

Since it is indicated [51] that UV-band shifts for iodine complexes (CTC-I) do not correlate with the solvent basicity B by a single equation, general for both n- and π-donors, of interest was to check whether a multiparameter equation is useful. As calculated for all substances in Table 15, the coefficient of correlation is low: $R = 0.918$. Successive elimination of data for mostly deviating points increases R: 0.942 for No. 19 (quinoline); 0.959 for No. 18 (pyridine); 0.966 for No. 25 (*o*-xylene).

Therefore, the following equation was deduced for 26 equations:

$$\Delta\nu\cdot 10^3 = 91.23 - 5{,}360\frac{n^2 - 1}{n^2 + 2} + 1{,}812\frac{\varepsilon - 1}{2\varepsilon + 1} - 1.66\delta^2 + 0.213B + 4.20E; \quad (70)$$

$R = 0.966$; $S = 138.4$;

$r_{01} = 0.711$; $r_{02} = 0.800$; $r_{03} = 0.681$; $r_{04} = 0.216$; $r_{05} = 0.784$.

Elimination of relatively insignificant parameters of specific solvation reduces the equation to the following three-parameter form:

$$\Delta\nu\cdot 10^3 = 20.2 - 5{,}709\frac{n^2-1}{n^2+2} + 1{,}943\frac{\varepsilon-1}{2\varepsilon+1} + 1.72\delta^2; \tag{71}$$

$R = 0.963$; $S = 138.1$.

In contrast with CTC-II, spectral band shift for iodine CTC-I is defined only by nonspecific solvation factors which conforms to nonionic type of CTC-I [51]. Elimination of separate, nonspecific solvation parameters disturbs the correlation:

$$R = 0.866 \text{ for } f(\varepsilon, \delta^2);\ R = 0.899 \text{ for } f(n, \delta^2);\ R = 0.904 \text{ for } f(n, \varepsilon). \tag{72}$$

Applicability of the Badger-Bauer rule to complexes with hydrogen bonds, i.e. the presence of satisfactory linear correlation between $\Delta\nu$ and ΔH of the complex formation, was under discussion for a long time. Whereas the presence of such linear dependence was argued in some works [98], Drago *et al.* [99, 100] concluded that sometimes failure of this rule is caused by inaccuracy of spectrophotometric determination of ΔH values. They showed that calorimetric techniques gave linear dependence of ΔH on $\Delta\nu$.

The works on applicability of the Badger-Bauer rule were performed for phenol complexes. For iodine complexes, linear dependence between spectrally determined ΔH_{J_2} and $\Delta\nu_{J_2}$ was not observed, despite some symbate property between ΔH and $\Delta\nu$ values (both in UV and visible regions). This is the reason for different dependence of $\Delta\nu$ on the donor properties compared with ΔH_{J_2}.

1.5. Application of LFE Equations to Generalize the Solvent Effect on Spectral Characteristics of Some CTC with Aromatic Hydrocarbons

Applicability of thė suggested approach was confirmed on some other examples. As shown above, location of CTC charge transfer band maximum and equilibrium constants of complex formation, determined from the Benesi-Hildebrand equation, noticeably depend on the environment. This phenomenon is explained [101] by different electron acceptor solvation in different solvents and formation of a charge transfer complex (CTC) with the acceptor. Another explanation is also acceptable – partial binding of acceptor molecules by active solvent reduces acceptor effective concentration that reflects upon complex formation parameters to be determined due to an original competition between the donor and the solvent for acceptor molecule. This event is known in chemical kinetics as slowing down a process in donor solvents [102]. Most likely, spectrally determined values result from the total effect of proceeding processes.

Nevertheless, the attempts of taking into account the effect caused by solvent universal interaction, defined by polarity and chargeability, on CTC parameters failed. For example, S.S. Diankov and V.V. Il'ev [105 – 108] performed systematic studies of Mac-Ray and Bakhshiev formula applicability to determination of a possible connection between location of the charge transfer band maximum (or transition energy $E_n = h\nu$) in UV spectrum of CTC from chloranil or tetracyanethylene and aromatic hydrocarbons with dielectric constant and refraction index of solvents. It is found that these formulae give satisfactory dependencies only for relatively inert solvents – aliphatic and alicyclic hydrocarbons, CS_2, and some hydrocarbons. However, data on CCl_4 or dichloroethane noticeably deviate from linear regression. Transition energies determined in aromatic hydrocarbons, alcohols, acetone and esters are higher and in $CHCl_3$, CH_2Cl_2, tetrachloroethane and acetonitrile lower than the calculated values. These deviations are explained by a contribution of specific interactions, which is confirmed by studies of the solvent effect on the transition energy in phenyl methyl ketone-iodine complex [106]. The absence of linear correlation between solvent polarity and CTC absorption band shift or the equilibrium constant for complexes from tetracyanethylene and aromatic hydrocarbons is confirmed [107]. Moreover, the authors state that "any solvent may not be accepted as an inert (inactive) one".

Table 16 shows data [103] on the solvent effect on the location of charge transfer band maximum for chloranil-pyrene complex.

Table 16. The solvent effect on charge transfer absorption frequencies ν in chloranil-pyrene complex [103]

No.	Solvent	ν, cm^{-1}
1.	Cyclohexane	16,900
2.	*n*-Hexane	17,100
3.	*n*-Heptane	17,000
4.	Ethanol	17,500
5.	*n*-Propanol	17,400
6.	*iso*-Propanol	17,300
7.	*n*-Butanol	17,100
8.	Dioxane	17,400
9.	Diethyl ether	17,400
10.	Acetone	17,600
11.	Acetic acid	17,000
12.	Methyl acetate	17,500
13.	Ethyl acetate	17,300
14.	Acetonitrile	16,700
15.	Benzene	16,800
16.	Dichloroethane	16,300
17.	Chloroform	16,100
18.	CCl_4	16,600

For all 18 solvents, data processing for band shift $\Delta\nu$ of chloranil-pyrene complex in relation to solution in cyclohexane (ν = 16,900 cm^{-1}) gave poor correlation (R = 0.875). Elimination of four, most deviating solvents (No. 4, 8, 9, 17) increases the coefficient of correlation R to a satisfactory level:

$$\Delta\nu = 1,615 - 3,247f(n) - 2,380f(\varepsilon) - 5.32\delta^2 + 6.189B - 12.14E; \quad (73)$$

$n = 14; R = 0.974; S = 114;$

$r_{01} = 0.400; r_{02} = 0.180; r_{03} = 0.072; r_{04} = 0.672; r_{05} = 0.186.$

Sequential elimination of parameters from the equation indicates insignificance of electrophilicity. Therefore the four-parameter equation describes the solvent effect on the charge transfer band shift quite accurately:

$$\Delta\nu = 1{,}748 - 3{,}655f(n) - 2{,}444f(\varepsilon) - 5.62\delta^2 + 5.920B; \quad (74)$$

$R = 0.96; S = 118.$

Elimination of any of the rest parameters makes the correlation worse to unsatisfactory ($R < 0.95$), and elimination of basicity disturbs it completely ($R < 0.5$). Therefore, the factor defining $\Delta\nu$ is nucleophilic solvation of a complex, which is confirmed by the highest paired coefficient of correlation for this parameter. Nucleophilic solvation of a complex or, most likely, binding of a part of chloranil by donor solvent increases the charge transfer energy; vice versa, nonspecific solvation of the complex reduces the energy similar to the increase of cohesive energy density, apparently, due to a "cage" formation around complex molecule. Possible effect of the cage effect of the solvent on CTC energy parameters is indicated [104] by increase of the complex formation constant with the polymer content for *o*-xylene-tetracyanethylene complex in binary isooctane-polyisobutylene mixtures. Reduction of the transition energy by nonspecific solvation correlates with the conclusions made in [103].

Data on the environment effect on transition energies E_{tr} are also generalized by LFE equations well for a series of tetracyanethylene or chloranil complexes [104 - 106].

All considered complexes are subject to general effect of nucleophilic solvation, which defines and increases the transition energy, and opposite effect of nonspecific solvation. Meanwhile, cohesive energy density and electrophilicity of the environment are insignificant factors [139].

Chloranil-hexamethyl benzene:

$$E_{tr} = 66.43 - 33.49f(n) - 3.53f(\varepsilon) - 0.017\delta^2 + 0.0051B + 0.133E; \quad (75)$$

$n = 11; R = 0.979; r_{OB} = 0.815; S = 0.63;$

$$E_{tr} = 63.41 - 25.38f(n) - 46.68f(\varepsilon) + 0.0040B + 0.068E; \quad (76)$$

$R = 0.954; S = 0.63.$

Chloranil-dimethylaniline:

$$E_{tr} = 50.43 - 17.97f(n) + 11.37f(\varepsilon) - 0.007\delta^2 - 0.0221B + 0.054E; \quad (77)$$

$n = 11; R = 0.975; r_{OE} = 0.620; S = 0.198;$

$E_{tr} = 49.87 - 15.56\mathbf{f}(n) - 0.007\delta^2 + 0.057E;$ (78)

$R = 0.960; S = 0.23.$

Since dimethylaniline is much stronger electron donor than hexamethyl benzene (basicities are 422 and 102, respectively [110]), the determining factor of the solvent effect on CTC energy is electrophilic solvation of the donor rather than nucleophilic solvation of the acceptor.

Tetracyanethylene-o-xelene:

$E_{tr} = 78.34 - 41.49\mathbf{f}(n) - 23.49\mathbf{f}(\varepsilon) + 0.069\delta^2 + 0.014B + 0.103E;$ (79)

$n = 8; R = 0.983; r_{OB} = 0.702; S = 0.90;$

$E_{tr} = 81.61 - 58.54\mathbf{f}(n) - 20.94\mathbf{f}(\varepsilon) + 0.077\delta^2 + 0.0127B;$ (80)

$R = 0.979; S = 0.98.$

Tetracyanethylene-naphthalene:

$E_{tr} = 75.23 - 125.80\mathbf{f}(n) - 13.63\mathbf{f}(\varepsilon) + 0.188\delta^2 + 0.0109B - 1.27E;$ (81)

$r_{OB} = 0.697; S = 0.54;$

$E_{tr} = 75.66 - 124.6\mathbf{f}(n) + 0.25\delta^2 + 0.0053B - 1.05E;$ (82)

$R = 0.967; S = 1.13.$

Tetracyanethylene-pyrene:

$E_{tr} = 47.21 - 18.86\mathbf{f}(n) - 4.77\mathbf{f}(\varepsilon) + 0.013\delta^2 + 0.0089B - 0.112E;$ (83)

$R = 0.938; r_{OB} = 0.773; S = 0.691;$

$E_{tr} = 48.27 - 24.51\mathbf{f}(n) - 8.89\mathbf{f}(\varepsilon) + 6.0038\delta^2 + 0.0089B - 0.0092E;$ (84)

$R = 0.929; S = 0.740.$

Iodine-acenaphthene:

$E_{tr} = 80.83 - 25.19\mathbf{f}(n) - 13.39\mathbf{f}(\varepsilon) + 0.014\delta^2 + 0.0356B - 0.0276E;$ (85)

$R = 0.997; r_{OB} = 0.960; S = 0.37;$

$E_{tr} = 81.30 - 25.95\mathbf{f}(n) - 10.10\mathbf{f}(\varepsilon) + 0.0348B;$ (86)

$R = 0.993; S = 0.48.$

Further elimination of parameters **f**(n) or **f**(ε) reduces R to 0.98 and 0.97, respectively.

For inactive solvents, transition energies in vacuum were determined by extrapolating the dependence E_{tr} = **f**(n, ε) [103 – 107], which were found close to experimental values. In the considered equation the corresponded role would be played by a_0 term which formally equals the transition energy in the environment without external impacts, at B, E and $\delta^2 = 0$ and n, ε = 1, e.g. in vacuum. Table 17 shows E_{tr} values in the gas phase, obtained by extrapolating spectral data by Mac-Ray and Bakhshiev equations, and experimental values.

Table 17. Experimental and theoretical CTC transition energies in vacuum [103 – 107] and free terms a_0 of correlation equations [139]

Complex	E_{exp}, kcal/mol	E_{theor}, kcal/mol		a_0
		Mac-Ray	Bakhshiev	
Chloranil-pyrene	-	54.3	52.5	46.5
Chloranil-hexachlorobenzene	-	61.92	-	66.4
Chloranil-dimethylaniline	-	52.92	-	50.43
Tetracyanethylene–*o*-xylene	75.26	75.33	73.08	78.34
Tetracyanethylene-naphthalene	60.35	-	-	75.23
Tetracyanethylene-pyrene	47.19	46.96	45.50	47.21
Iodine-acenaphthene	-	84.12	-	80.93

Comparison of the data with the authors' a_0 in equations indicates a satisfactory coincidence for tetracyanethylene-pyrene and tetracyanethylene-*o*-xylene complexes, whereas tetracyanethylene-naphthalene complex produces high deviation. The rest complexes possess no experimental data, and a_0 values determined in [103 – 107] differ from calculated ones by 3 – 4 kcal/mol. Such deviations are caused by inclusion of highly solvated media into calculation.

Data on the environment effect on forming complexes from anthracene, acenaphthene, 9-methylanthracene and 2,6-dimethoxy naphthalene with tetrachlorophthalic and 3,5-dinitrophthalic anhydrides also confirmed applicability of the LFE equations [108 – 111]. The authors [109, 110] considered 32 solvents. In the current monograph, we use only 24 solvents, for which necessary data are available in the literature. Unfortunately, generalization by the five-parameter equation gives poor correlation (R = 0.77); therefore, three solvents with most highly deviating parameters (pyridine, dioxane and cyclohexane) were eliminated. This increased R to 0.965, and subsequent elimination of chloroform – to R = 0.975. As a result, the following equation was deduced:

$$\nu_{max} \cdot 10^3 = 25.70 - 3.39\frac{n^2 - 1}{n^2 + 2} - 2.19\frac{\varepsilon - 1}{2\varepsilon + 1} +$$

$$+ 0.03\delta^2 + 0.00318B + 0.027E; \quad (87)$$

$n = 20$; $R = 0.975$; $S = 0.10$.

$r_{01} = 0.55$; $r_{02} = 0.37$; $r_{03} = 0.38$; $r_{04} = 0.89$; $r_{05} = 0.23$.

Nonspecific solvation reduces the charge transfer (CT) energy and, therefore, stabilizes the complex; specific solvation increase CT energy due to competition between donor and solvent for acceptor molecule. The solvent ability to self-association also increases this value.

In accordance with the paired coefficients of correlation nucleophilic solvation in the most significant factor affecting $\Delta\nu$: its elimination reduced R to 0.63. The effects of the rest parameters are less significant, but also noticeably contribute into CT energy: at their subsequent elimination R varies within 0.9 – 0.94, which is poor correlation. Electrophilic solvation is the only insignificant factor:

$$\nu\cdot 10^3 = 25.88 - 4.12\frac{n^2-1}{n^2+2} - 2.25\frac{\varepsilon-1}{2\varepsilon+1} + 0.004\,\delta^2 + 0.03B; \quad (88)$$

$R = 0.973$; $S = 0.10$.

Linear dependence between ν_{max} and solvent basicity was also observed [108, 109], however, for 12 *n*-donor solvents only. Data on dioxane, benzonitirile and benzaldehyde fit another line, and data on π-donors fully deviate from this dependence. It is suggested to subdivide solvents into several classes (a – halohydrocarbons; b – aromatic hydrocarbons; c – aromatic derivatives and pyridine; d – aliphatic *n*-donors) due to differences in the main and excited states of solvated solvents, because correlation dependencies may be built within the same class of solvents only. However, calculations performed indicate satisfactory generalization of various effects of the environment (specific and nonspecific solvation, cage effect), impacting CT energy of the complex, on the basis of the LFE principle and quantitative description by linear multiparameter equations.

Satisfactory results were also obtained for description of the environment effect on CT band shifts in other complexes of phthalic anhydrides [110]. Solvent basicity (ability to nucleophilic solvation) is always determining factor for the complexes, signs and significances of separate regression terms are preserved (except for 9-methylanthracene complex with tetrachlorophthalic anhydride). This indicates similar chemical properties of solvation for all cases [111].

The absence of correlation between $\Delta\nu$ and solvent polarity and the presence of a definite dependence on chargeability are indicated [108 – 110]. These qualitative conclusions are confirmed in digits: the paired coefficient of correlation for $\frac{\varepsilon-1}{2\varepsilon+1}$ term is low, therefore, it is insignificant, whereas the contribution of chargeability is noticeable for CT band shift.

Shifts in UV-spectra of some compounds in organic solvents are also satisfactorily described by multiparameter equations, and solvation processes in these solvents are interpretable. As an example, we present data [112] on absorption maximum shifts in UV-spectra of 4-dialkylamine-α-cyanostilbene-4-diazonium tetrafluoroborate in 50 solvents. In this case, the environment effect may not be described by one-parameter dependencies on basicity (β) or bipolarity (π^x) of the Kamlet-Abraham-Taft equation. It is explained by possible existence of various forms of diazo-compounds with respect to the solvent type – ionic, ionic associate or ionic dipole. These data are properly generalized by the Koppel-Palm three-parameter equation, where basicity increases and nonspecific solvation decreases ν [113].

Numerous examples of application of multiparameter Koppel-Palm, Kamlet-Abraham-Taft and other equations to generalize and interpret shifts in UV, NMR and IR spectra are shown in [114].

Therefore, mathematical analysis of the experimental data induces the following conclusions:

1. The energies of charge transfer band spectral transition in CTC depend on the type of environment, the determining effect being caused by specific solvation of electron donor or acceptor by solvent, which increases energy.
2. Nonspecific (universal) interaction with the solvent stabilizes the formed complex and reduces CT energy.
3. The dependence of energy of transfer frequency on physical and chemical properties of the environment is mostly described by linear five-parameter equation of linear free energies, solvent ability to specific and nonspecific solvation, and cohesive energy density of the solvents.

2. Generalization of Heats of Mixing by LFE Equation

The environment effect (independently of the fact, if it is a partner in the complex formation or not) on the complex formation parameters conforms to the ideas about the role of solvation effects, quantitatively defined by the LFE principles, but for heats of mixing, determined in concentrated solutions, this approach is ineligible. On the contrary, the heats of mixing for some acceptors and donors are suggested to be electron donor property: with $CHCl_3$ [115, 116], $SbCl_5$ (*DN*) [5], SO_3 [117], as well as dissolution heats *KJ* [118, 119]. However, as mentioned above, these different scales of the "donor property" are mutually disproportional. The more so, as the solvent with calorimetric interactions is replaced, for example, from dichloroethane to benzene, values of interaction enthalpies for *DN* change abruptly. Nevertheless, processing of a broad material on enthalpies of mixing outlines a satisfactory correlation of heats of mixing of various acceptors with several donors with donor properties, obtained by multiparameter LFE equations. Such correlation was performed by comparing basicity scales *DN* and *B* [120], which was found moderate for them personally:

$$DN = -0.69 + 0.088B;\ n = 41;\ R = 0.958;\ S = 3.5. \quad (89)$$

It significantly increases as nonspecific solvation, induced by chargeability, is taken into account. In this case, polarity and cohesive energy density of the donor are insignificant.

Such multiparameter dependencies were also obtained for calorimetric effects of mixing of other acceptors – bromine, $SbCl_3$, $AsCl_3$, $AlBr_3$, $CHCl_3$. Here of the determining importance is the acid-base interaction, i.e. donor basicity factor. However, as Lewis acidity of the acceptor decreases, the value of other solvation processes increases.

Processing of data on enthalpies of bromine mixing with several substances [121 - 123] gave the following dependencies:

$$\Delta H_{Br_2} = -6.25 + 38.53f(n) + 109.01f(\varepsilon) - 0.0082\delta^2 - 2.55B + 5.69E; \quad (90)$$

$n = 21$; $R = 0.976$; $S = 3.0$;

$$\Delta H_{Br_2} = -10.0 + 60.99f(n) - 2.44B + 2.88E; \quad (91)$$

$R = 0.963$; $S = 3.5$.

Expectedly, the ability of the second component to electrophilic solvation reduces its ability to interaction with acid (by Lewis) bromine molecule and, correspondingly, the heat effect of mixing. Enthalpies of pyridine-bromine (ΔH = -103.664 kJ/mol) and dioxane-bromine (ΔH = -40.96 kJ/mol) ionic complex formation were used in calculations. As ΔH = -103.66 kJ/mol is replaced by -58.52 kJ/mol, determined for solid molecular complex [139], correlation is noticeably disturbed ($R = 0.919$). Meanwhile, direct correlation between ΔH and DN is rather low:

$\Delta H = 6.52 - (2.72 \pm 0.27)DN$; $R = 0.928$.

Contrary to calorimetric data, spectrally obtained enthalpies of bromine complex formation with donors are significantly contributed by chargeability. For instance, the dependence obtained is not fit with enthalpies of bromine mixing with substituted pyridines in dichloroethane [62], falling within the range of 33.44 - 45.98 kJ/mol, as well as in dioxane – 8.36, tetrahydrofuran – 14.63, dimethyl sulfide 37.20 [54, p. 311], ethyl ether – 22.99 [124], and N,N-diethyl acetamide – 6.688 kJ/mol [125], determined using a spectrophotometry technique. Correspondingly, there is no good correlation between calorimetric enthalpies of donor mixing with bromine and spectral enthalpies of complex formation from the same donors and iodine. The following dependence was deduced for 10 solvents:

$$\Delta H_{Br_2} = -5.49 + (3.62 \pm 0.50)\Delta H_{J_2}; \ R = 0.933; \ S = 4.9. \quad (92)$$

Oscillation frequencies for ν_{Br}-ν_{Br} band in the Raman spectrum were determined [126] at 296 K for 22 solvents. The authors failed to associate the frequency shift with polarity $\frac{\varepsilon - 1}{2\varepsilon + 1}$ and concluded about the determining effect of the donor-acceptor interaction for this value. Though correlation between $\nu_{Br\text{-}Br}$ and Palm's basicity B is much higher than with polarity, for which $R = 0.250$, it is poor, too: $R = 0.772$ for 19 solvents. For five-parameter equation, total coefficient of correlation increases to 0.874. After elimination of data for three mostly deviating solvents (ethanol, benzene and fluorobenzene), the equation with acceptable correlation was obtained [120]:

$$\nu_{Br\text{-}Br} = 329.0 - 54.83f(n) + 4.12f(\varepsilon) - 0.0071\,\delta^2 - 0.064B + 0.103E; \quad (93)$$

$R = 0.957$; $S = 1.56$.

Elimination of cohesive energy density and electrophilicity, which are insignificant parameters, adequate reduced equations were obtained:

$$\nu_{Br\text{-}Br} = 328.7 - 53.65f(\text{n}) + 260f(\varepsilon) - 0.063B; \quad (94)$$

$R = 0.949$; $S = 1.56$

or

$$\nu_{Br\text{-}Br} = 329.5 - 53.39\,f(\text{n}) - 0.052B; \quad (95)$$

$R = 0.947$; $S = 1.58$.

Thus, similar to enthalpy of mixing, donor basicity is the factor determining spectral band shift, and a noticeable effect is caused by nonspecific solvation, first of all, solvent chargeability, which is typical of optical processes, for example, for solvent effect on UV absorption of iodine [127]. It is clear that for seven general points, linear dependence between $\nu_{Br\text{-}Br}$ and ΔH_{Br_2} is observed only for spectrally measured dioxane enthalpy value equal 8.36 kJ/mol. Therefore, bromine mixing with organic donor is accompanied by the donor-acceptor interaction. Other solvation effects make a significant contribution, though it is different for calorimetric and spectral effects. Satisfactory correlation between the studied function and solvent parameters may be obtained only at taking into account all solvation types via linear five-parameter energy equation.

Nonspecific solvation effects become more important at acceptor force and heat effect absolute value decrease. Data on $AsCl_3$, $SbCl_3$ and $AlBr_3$ dissolution heats were generalized [128]. In all cases, the five-parameter equation gives good correlation between enthalpy of mixing (dissolution) and solvent properties. However, the effect of specific, acid-base interaction regularly decreases: $r_{OB} = 0.952$ for $SbCl_3$, 0.917 for $AsCl_3$, and only 0.775 for $AlBr_3$. Note also a significant contribution of donor polarity into the total effect, especially in the case of $AlBr_3$. This is associated with Al_2Br_6 dissociation in more polar environment and its transition to a monomeric form.

Expectedly, the reverse dependence is observed for ethylene diamine [129]. Heats of mixing of this strongest electron donor ($DN = 50 \pm 3$) [130], $B_{PhOH} = 560$ [131] are caused by electrophilicity of the second reagent. One-parameter data processing gives the following results [130]:

$$\Delta H = 147.97 - (0.869 \pm 1.354)E_T;\ n = 15;\ R = 0.884. \quad (96)$$

$$\Delta H = 29.26 - (1.906 \pm 0.209)AN;\ n = 11;\ R = 0.951. \quad (97)$$

Much better generalization is obtained at consideration of other solvent parameters:

$$\Delta H = 104.1 - 397.96f(n) + 0.004\delta^2 - 1.59E; \quad (98)$$

$$n = 15;\ R = 0.957;\ S = 2.58.$$

As a consequence of high correlation between enthalpies of ethylene diamine mixing with various substances and their electrophilicity (acidity), an accessory electrophilicity scale may be composed on this basis.

The contribution of nonspecific interactions to total calorimetric effect of mixing dramatically increases with weakening of acid-base, specific interactions. A broad experimental material [132] on heats of chloroform mixing with more than 60 substances was studied in detail [133]. However, processing of the whole data array, i.e. points 1 – 45 in Table 18, which possess all required parameters, did not give good correlation: $R = 0.693$.

It is shown that obtaining good linear dependence of C-H bond deformation vibration shift in chloroform on the solvent properties requires separate consideration of proton and aprotic (associated and unassociated) solvents. Such separation into two groups is also currently performed, where hydroxyl-containing compounds with hydrogen bonds and tertiary amines able to associate by means of *n*-electron pairs are related to the class of associated solvents. The correlation analysis confirms rightfulness of such separation.

The dependence of heats of chloroform mixing with unassociated solvents (No. 1 – 33) on their physical and chemical properties is described by the five-parameter equation (R = 0.956) at paired coefficient of correlation for basicity $r = 0.909$. Solvents No. 34 – 41 were not included in the calculation, because they had no data on E_T. To increase adequacy of the equation, mostly deviating data for solvents No. 32 and 26 were eliminated:

$$-\Delta H_{CHCl_3} = -6{,}109.7 + (611 \pm 2{,}308)f(n) - (5{,}626 \pm 1{,}769)f(\varepsilon) -$$

$$- (4.8 \pm 1.1)\delta^2 + (12.5 \pm 0.7)B + (213 \pm 55)E_T; \quad (99)$$

$$n = 31;\ R = 0.977;\ S = 329.7.$$

Since $r_{OB} = 0.959$, the main contribution into ΔH_{CHCl_3} is made by basicity. The rest parameters are insignificant. Subsequent elimination of nonspecific solvation parameters, e.g. polarity and chargeability, does not deteriorate total coefficient of correlation, and dependencies of heats of chloroform mixing with unassociated solvents are well described by the three-parameter equation:

$$-\Delta H_{CHCl_3} = -778 - (3.35 \pm 1.18)\delta^2 + (11.55 \pm 0.77)B +$$

$$+ (50.7 \pm 32.3)E; \quad (100)$$

Table 18. Heats of mixing of organic substances with chloroform

No.	Solvent	$-\Delta H$, J/mol
1.	Tetrachloromethane	-232.4
2.	Dichloromethane	-18.0
3.	*o*-Dichlorobenzene	0
4.	Bromobenzene	142.1
5.	Nitromethane	133.8
6	*α*-Chloronaphthalene	154.7
7.	Chlorobenzene	167.2
8.	Benzene	451.4
9.	Nitrobenzene	493.2
10.	Dichloroethane	0
11.	Toluene	656.3
12.	*m*-Xylene	844.4
13.	*o*-Xylene	865.3
14.	*p*-Xylene	886.2
15.	Mesitylene	902.9
16.	Acetic anhydride	1,280.8
17.	Benzonitrile	1,241.5
18.	Methyl acetate	1,546.6
19.	Butyrolacton	1,584.2
20.	Acetone	1,897.7
21.	Ethyl acetate	1,935.3
22.	Diethyl carbonate	2,186.1
23.	Dioxane	2,215.4
24.	Cyclohexanone	2,407.7
25.	Diethyl ether	2,708.6
26.	Dimethyl sulfoxide	2,746.3
27.	Tetrahydrofuran	2,804.8
28.	Diisopropyl ester	2,834.0
29.	Dibutyl ether	3,093.2
30.	Dimethyl acetamide	3,845.6
31.	Triethyl phosphate	4,598
32	Dimethaoxyethane	4,932.4
33.	Tributyl phosphate	5,183.2
34.	Benzoyl chloride	349.0
35.	Thionyl chloride	501.6
36.	Sulforyl chloride	693.9
37.	Diethyl sulfite	827.6
38.	Methyl propionate	1,818.3
39.	Phosphoryl chloride	1,977.1
40.	Tetrahydropyran	2,508
41.	Trimethylene oxide	2,938.5
42.	Isobutanol	-774.0
43.	Acetic acid	-576.8

Table 18. Heats of mixing of organic substances with chloroform (Continued)

No.	Solvent	-ΔH, J/mol
44.	Aniline	-192.3
45.	*n*-Propanol	-129.6
46.	Ethanol	246.6
47.	Methanol	346.9
48.	Dimethylaniline	1,019.9
49.	Pyridine	2,257.2
50.	Triethylamine	4,025.3
51.	Isoamyl alcohol	-418
52.	Pyrrole	6,06.1

$R = 0.966$; $S = 385$.

Thus, besides solvent basicity determining the heat effect, heats of chloroform mixing with aprotic solvents are slightly affected by self-association of donor that reduces the heat effect. Note that maximal deviation of data on dimethoxyethane may be explained by presence of two donor centers. This gives an opportunity to form complexes with chloroform of 1:1 and 1:2 ratio. Simultaneous insignificant electrophilic solvation of chloroform is also possible, most likely, due to weak interaction of solvent molecules with chlorine atom electrons in $CHCl_3$. Elimination of E_T parameter slightly reduces the coefficient of correlation:

$$-\Delta H_{CHCl_3} = 496.6 - (1.82 \pm 0.69)\delta^2 + (12.2 \pm 0.7)B; \quad (101)$$

$n = 31$; $R = 0.966$; $S = 385$.

This two-parameter equation allows checking conformity with total calculation for solvents No. 31 – 41, not included in the primary calculation by the five-parameter equation due to the absence of data on E_T and ε parameters. The majority of these data conform to the linear dependence for tertiary amines No. 48 – 50.

The situation is different for associated solvents No. 42 - 50. Here correlation between enthalpies and solvent properties is described by the five-parameter equation ($R = 0.977$), but correlation with basicity is lower ($r_{OB} = 0.926$), though even in this case this is the determining factor. The second significant factor is chargeability of the environment, and cohesive energy density and electrophilic solvation are insignificant:

$$-\Delta H_{CHCl_3} = 192.6 - (11{,}090 \pm 3{,}280)f(n) + (10.31 \pm 0.97)B; \quad (102)$$

$n = 9$; $R = 0.975$; $S = 395$.

Note also that Tamres recommended ΔH_{CHCl_3} values [133] as substance basicity characteristics. However, the data obtained display no general dependence on Palm's basicity

B. In the literature, three linear correlations for various classes of substances are reported: ethers [134], amines [115] and aromatic hydrocarbons [133], where some substances deviate from linearity [134]. Contrary to one-parameter dependence between *DN* (i.e. heat of mixing with $SbCl_5$) and Palm's basicity of substances, for chloroform nonspecific interactions should be taken into account, and unassociated and associated substances must be considered separately. This is so, because for $SbCl_5$ heat effects equal 4.18 – 125.4 kJ/mol, and for $CHCl_3$ – oly 0.01 – 0.5 kJ/mol.

As shown above, such dividing of solvents into two groups of associated and unassociated solvents is also necessary at data generalization on the environment effect on shifts of C-H bond deformation vibration band in $CHCl_3$ ($\delta = 1{,}210$ cm^{-1}) or C-D valence vibrations in $CDCl_3$ ($\nu = 2{,}252$ cm^{-1}).

Heat effects of mixing with aniline and acetone are much more sensitive to impact of perturbing factors. Aniline possesses properties of both donors and acceptors, because it is able to form hydrogen bond (B = 346, E = 6.2); for acetone B = 224, E = 2.1 [135]. Complexity of interactions in the systems with these substances is confirmed by the fact that, despite stronger donors (acceptors), maximum of the heat effect may be observed at ratios of components different from 1:1. This indicates a possibility of forming compounds of another composition (aniline + MeCN, MeOH, Et_2O; acetone + water, $HCONH_2$, $CHCl_3$). Nevertheless, it was found possible to generalize data on heats of aniline mixing (except for systems with asymmetrical dependencies) by the following equation:

$$\Delta H = -423.85 + 4{,}034f(n) - 5{,}660f(\varepsilon) + 9.51\delta^2 - 0.02B - 185.6E; \quad (103)$$

$n = 15$; $R = 0.973$; $S = 67$.

Chargeability is found insignificant, because at its elimination R is reduces to 0.970. Expectedly, self-association of solvents suppresses their interaction with aniline (positive regression coefficient at δ^2), and specific solvation increases heat effect of the interaction. Taking into account complex type of donor and acceptor effects, and low heat effects of mixing (maximally, -1,672 J/mol for pyridine), the deduced equation should be accepted.

Similar picture is observed for acetone. Moderate donor type of this compound gives mostly positive enthalpies of its mixing with other donors, which fall below 1.67 J/mol. Adequate generalization of these data requires elimination of systems with asymmetrical maximum and data on associated substances – water, alcohols, formamide, and aniline. Similar to the previous case, the heat effect of mixing depends on different types of component interaction.

Applicability of multiparameter LFE equations to data generalization on enthalpies of mixing was also proved for inorganic acid (N_2O_4, SO_3) and proton acid (*tert*-butyl hydroperoxide) anhydrides [136]. Such approach using Kamlet-Taft solvent parameters was used [137, 138] for generalizing calorimetric data on mixing a series of substances with alcohols and acetone.

Thus, there is no doubt that LFE relations are effective for description of complex formation between electron donors and acceptors by both strong and weak hydrogen bonds and charge transfer. The use of linear, multiparameter equations gives positive results for correlating spectral shifts (IR, UV, ESR), complex formation constants and enthalpies, calorimetric of Lewis acid (electron acceptor) mixing with donors, and describing the effect of the third component (the environment) on CTC characteristics.

References

[1] Fowler F.W. and Katritzky A.R., *J. Chem. Soc.*, 1971, №3, pp. 460 - 469.

[2] Koppel I.A. and Palm V.A., In: *Advances In Linear Free Energy Relationships*, Ed. N.B. Chapman and A.J. Shorter, London, Plenum Press, 1973, pp. 203 - 281.

[3] Gutmann V., *Chemistry Of Coordination Compounds In Non-aqueous Solutions*, Moscow, Mir, 1972, pp. 22 – 50. (Rus)

[4] Drago R.S., In: *Structure And Bonding*, 1973, No. 15, pp. 73 - 139.

[5] Gutmann V. Abhandl., *Braunschw. Wiss. Geselsch.*, 1977, Bd. 28, S. 7 - 22.

[6] Iohansen A.V., *Teor. Eksper. Khim.*, 1971, vol. 7(3), pp. 302 - 311. (Rus)

[7] Drago R.S., Parr L.B., and Chamberlain C.S., *J. Am. Chem. Soc.*, 1977, vol. 99(10), pp. 3203 - 3209.

[8] Pearson R.C., *J. Amer. Chem. Soc.*, 1963, vol. 85(22), pp. 3533 - 3539.

[9] Terentiev V.A., *Thermodynamics Of Donor-Acceptor Relationship*, Izd. Saratovskogo Univ., 1981, 278 p. (Rus)

[10] Sadovsky O.S., Panchenko B.V., Oleinik N.M., and Dadali V.L., *Ukr. Khim. Zh.*, 1979, vol. 45(7), pp. 641 - 647. (Rus)

[11] Krygowski T.M. and Fawcett W.R., *J. Amer. Chem. Soc.*, 1975, vol. 97(8), pp. 2143 - 2148.

[12] Fawctet W.R. and Krygowski T.M., *Canad. J. Chem.*, 1976, vol. 54(20), pp. 3283 – 3292.

[13] Yatsimirovskaya N.T. and Busev A.I., *Zh. Neorg. Khim.*, 1979, vol. 24(12), pp. 3183 - 3187. (Rus)

[14] Mayer U., *Monatsch. Chem.*, 1978, Bd. 109(2), S. 421 - 433.

[15] Swain C.G., Swain M,C., Powell A.C., and Alunnt S., *J. Amer. Chem. Soc.*, 1983, vol. 105(3), pp. 502 - 513.

[16] Szczepaniak K., In: *Wiazanie wodorowe*, Warszawa: PWN, 1969, st. 120 - 144.

[17] Iogansen A.V., In: *Hydrogen Bond*, Moscow, Nauka, 1981, pp. 112 - 152. (Rus)

[18] Powell D.L., West R., Whatley L.S. et.al., *Spectrochim. Acta*, 1964, vol. 20(6), pp. 984 - 991.

[19] Horak M., Polakova J., and Jakoubkova M., *Collect. Czech. Chem. Commun.*, 1966, vol. 31(2), pp. 622 - 633.

[20] Bellamy L.J., Hallam H.E., and Williams R.L., *J. Chem. Soc.*, 1958, No. 4, pp. 1120 - 1127.

[21] Bellamy L.J. and Pace W.J., *Spectrochim. Acta*, 1969, vol. 25A(2), pp. 319 - 328.

[22] Koppel I.A. and Payu A.I., *Reakts. Spos. Org. Soed.*, 1974, vol. II(1), pp. 121 - 138. (Rus)

[23] Bekarek V. and Knopova M., *Collect. Chech. Commun.*, 1977, vol. 42(6), pp. 1976 - 1980.

[24] Makitra R.G. and Pirig Ya.N., *Reakts. Spos. Org. Soed.*, 1980, vol. 18(2), pp. 184 - 205. (Rus)

[25] Makitra R.G., Zhukovsky V.Ya., Pirig Ya.N., and Chernyak B.I., *Zh. Prikl. Spektr.*, 1982, vol. 36(6), pp. 962 - 968. (Rus)

[26] Zhukovsky V.Ya., Pirig Ya.N., and Makitra R.G., *ZhOKh*, 1981, vol. 55(7), pp. 1833 - 1835. (Rus)

[27] Bitman G.L., Karasev Yu.V., and Elemberg M.E., *ZhOKh*, 1974, vol. 44(6), pp. 1361 - 1368. (Rus)

[28] Rossarie J., Gallas J.P., Binet C., and Romanet R., *J. Chem. Phys.*, 1977, vol. 74(2), pp. 188 - 196.

[29] Ioganssen A.V., Kurkchi G.A., Surman V.M., and Glazunov V.P., *Zh. Prikl. Spektr.*, 1980, vol. 33(2), pp. 302 - 308. (Rus)

[30] Purcell K.F., Stikeleather J.A., and Brunn S.D., *J. Amer. Chem. Soc.*, 1969, vol. 91(15), pp. 4019 - 4027.

[31] Ioganssen A.V., Rassadin E.V., and Sorokina N.P., *Zh. Prikl. Spektr.*, 1980, vol. 32, pp. 1089 - 1095. (Rus)

[32] Rossarie J., Gallas J.P., Binet C., and Romanet R., *J. Chim. Phys.*, 1977, vol. 74(2), pp. 197 - 201.

[33] Rossarie J., Gallas J.P., Binet C., and Romanet R., *J. Chim. Phys.*, 1977, vol. 74(2), pp. 202 - 206.

[34] Abd-el-Kerim A.L,Bassyoni A.H., and Shabana A.A., *Appl. Spectrosc.*, 1977, vol. 31(2), pp. 131 - 134.

[35] Makitra R.G. and Pirig Ya.N., *Zh. Prikl. Spektr.*, 1980, vol. 33(6), pp. 1071 - 1076. (Rus)

[36] Perelyigin I.S. and Afanas'eva I.M., *Zh. Prikl. Spektr.*, 1973, vol. 30(4), pp. 676 - 680. (Rus)

[37] Phutela R. and Arora P.S., *Z. Phys. Chem.*, 1976, Bd. 257(5), S. 945 - 949.

[38] Gordy W., *J. Amer. Chem. Soc.*, 1938, vol. 60(3), pp. 605 - 612.

[39] Gushchin V.A., *Optika I Spektroskopia*, 1964, vol. 17(3), pp. 385 - 390. (Rus)

[40] Mierzecki R., In: *Wiazanie wodorowe*, Warszawa: PWN, 1969, str. 145 - 174.

[41] Dervil E., 'Etude par spectroscopigue infrarouge', *Thes. Doct. Sci.*, Paris Univ., 1974; RZhKhimia, 1976, 20 B, p. 194. (Rus)

[42] Imaishi H., *J. Sci. Hiroshima Univ.*, Ser.A., 1975, vol. 39(3), pp. 337 - 355.

[43] Makitra R.G., Pirig Ya.N., Tsikanchuk Ya.M., and Zhukovsky V.Ya., *Ukr. Khim. Zh.*, 1980, vol. 46B(7), pp. 729 – 734. (Rus)

[44] Khare B.N., Mitra S.S., and Lengyel D., *J. Chem. Phys.*, 1967, vol. 47(12), pp. 5175 - 5179.

[45] Lord R.C., Nolin B., and Stidham H.D., *J. Amer. Chem. Soc.*, 1955, vol. 77(10), pp. 1365 - 1368.

[46] Makitra R.G., Kivelyuk R.V., and Pirig Ya.N., *Koord. Khim.*, 1982, vol. 8(5), pp. 605 - 606. (Rus)

[47] Bertran J.F. and Rodriguez M., *Org. Magn. Res.*, 1979, vol. 12(2), pp. 92 - 94.

[48] Makitra R.G., Protsailo G.A., Boldeskul I.E., and Rozhkova Z.Z., *Zh. Prikl. Spektr.*,1982, vol. 37(5), pp. 784 - 790. (Rus)

[49] Eyman D.E. and Drago R.S., *J. Amer. Chem. Soc.*, 1966., vol. 88(7), pp. 1617 - 1620.

[50] Benesi H.A. and Hildebrand J.H., *J. Am. Chem. Soc.*, vol. 71(8), pp. 2703 – 2707.

[51] Gorodisskii V.A. and Morachevskii A.A., *ZhFKh*, 1977, vol. 51(6), pp. 1313 - 1317. (Rus)

[52] Endryus L. and Keefer R., *Molecular Complexes In Organic Chemistry*, Moscow, Mir, 1967, 206 p. (Rus)

[53] Briegleb E., *Elektronen-Donator Komplexe*, Berlin: 1961, Springer Verl., 320 S.

[54] Gurianova E.I., Gol'dstein I.P., and Romm I.P., *Donor-Acceptor Relationships*, Moscow, Khimia, 1973, 400 p. (Rus)

[55] Terenin A.I., *Uspekhi Khimii*, 1955, vol. 24(2), pp. 121 - 162. (Rus)

[56] Mullikan R.S., *J. Amer. Chem. Soc.*, 1950, vol. 72(1), pp. 600 - 608.

[57] Keefer R.M. and Andrews L.J., *J. Amer. Chem. Soc.*, 1950, vol. 72(11), pp. 5170 - 5173.

[58] Borod'ko Yu.G. and Syirkin Ya.K., *Zh. Strukt. Khim.*, 1961, vol. 2(4), pp. 480 - 483. (Rus)

[59] Mullikan R.S., *J. Amer. Chem. Soc.*, 1952, vol. 74, pp. 811 - 824.

[60] Gazizov I.G. and Chmutova G.A., *ZhOKh*, 1980, vol. 50(9), pp. 2063 - 2070. (Rus)

[61] Sergeeva G.B., Smirnov V.V., Myishyakov M.N., and Borisova L.M., *Khim. Fiz.*, Iss. 9, pp. 1193 - 1196. (Rus)

[62] Tilguin B. and Lambert, *Z. Phys. Chem.* (DDR), 1973, Bd. 84(1-4), S. 84 - 91.

[63] Cheng-Zhang Yiang, Fu-Juan H.S.U., and Fried V., *J. Chem. Eng. Data*, 1973, vol. 28(1), pp. 75 – 78.

[64] Lux F., Klein O., and Paetzold R., *Z. Chem.*, 1978, Bd. 18(3), S. 107 - 108.

[65] Kirilicheva V.G., Kuznetsov V.A., and Egorochkin A.N., *Thes. Rep. II All-Russian Conf. Thermodynamics Of Organic Compounds*, 1975, Gorky, 1976, p. 36. (Rus)

[66] Tsarevskaya M.N., Zadorozhnaya N.T. and Yakimova Z.I., *ZhFKh*, 1982, vol. 56(10), pp. 2525 - 2526. (Rus)

[67] Drago R.S. and Weyland B.R., *J. Amer. Chem. Soc.*, 1965, vol. 87(16), pp. 3571 - 3572.

[68] Panchenko B.V., Oleinik N.M., Sadovsky Yu.S., Dadali V.A., and Litvinenko A.M., *Reakts. Spos. Org. Soed.*, 1980, vol. 17(I-61), pp. 69 - 89. (Rus)

[69] Sambhi M.S., *J. Mol. Struct.*, 1975, vol. 27(1), pp. 55 - 58.

[70] Perov A.N., Piterskaya I.V., and Bakhshiev N.G., *Optika I Spektroskopia*, 1971, vol. 30(5), pp. 872 - 876. (Rus)

[71] Fialkov Yu.Ya. and Borovikov V.Ya., *The Effect Of Solvent On Donor-Acceptor Interactions*, Dep. VINITI No. 1022-75, 14 p.; No. 1892-75, 12 p., No. 2887-75, 10 p.; No. 1073-78, 10 p. (Rus)

[72] Uruska I. and Szpakowska M., *Bull. Acad. Pol. Sci., Ser. Chem.*, 1972, vol. 20(5), pp. 441 - 458.

[73] Huong P.V., Platzer N., and Josien M.L., *J. Amer. Chem. Soc.*, 1969, vol. 91(13), pp. 6669 - 6670.

[74] McKinney W. and Popov A.I., *J. Am. Chem. Soc.*, 1969, vol. 91(19), pp. 5215 - 5219.

[75] Bist H.D. and Pearson W.B., *J. Phys. Chem.*, 1967, vol. 71(8), pp. 2750 - 2752.

[76] Uruska I., Kamienska E., and Szpakowska M., *Bull. Acad. Pol. Sci., Ser. Chem.*, 1972, vol. 20(8), pp. 805 - 812.

[77] Betrand G.L. Oyler D.E., Eichelbaum U.G., and Hepler L.G., *Thermochim. Acta*, 1973, vol. 7(2), pp. 87 - 93.

[78] Kiselev V.D., Mavrin G.V., Korshin E.E., and Konovalov A.I., *ZhOKh*, vol. 50(5), pp. 1135 - 1138. (Rus)

[79] Hastings S.H., Franklin J.L., Schiller J.C., and Matsen F.A., *J. Amer. Chem. Soc.*, 1953, vol. 75(6), pp. 2900 - 2906.

[80] Syitin M.S., *ZhFKh*,1974, vol. 48(9), pp. 228 - 229. (Rus)

[81] Fialkov Yu.Ya. and Borovikov A.Ya., Ukr. Khim. Zh., 1976, vol. 42(8), pp. 808 - 810. (Rus)

[82] Tsikanchuk Ya.M., Makitra R.G., and Luk'yanets V.M., *ZhOKh*, 1977, vol. 47(8), pp. 1690 - 1694. (Rus)

[83] Meyer M., Meyer R., Metzger J., and Peneloux A., *J. Chem. Phys. Physicochem. Biol.*, 1971, vol. 68(3), pp. 413 - 416.

[84] Detellier C. and Laszlo P., *J. Phys. Chem.*, 1976, vol. 80(22), pp. 2503 - 2506.

[85] Homer J. and Dudley A.R., *J. Chem. Soc.*, 1973, Faraday Trans., Part 1, No. 10, pp. 1995 - 2005.

[86] Bilobrov V.M. and Shchupach V.I., *ZhOKh*, 1982, vol. 52(4), pp. 729 - 734. (Rus)

[87] Bilobrov V.M. and Shchupach V.I., *ZhOKh*, 1982, vol. 52(4), pp. 735 - 739. (Rus)

[88] Makitra R.G. and Tsikanchuk Ya.M., *Reakts. Spos. Org. Soed.*, 1979, vol. 16(4), pp. 506 - 511. (Rus)

[89] Voigt E.M., *J. Phys. Chem.*, 1968, vol. 72(9), pp. 3360 - 3365.

[90] Tamres M., *J. Phys. Chem.*, 1961, vol. 65(4), pp. 654 - 664.

[91] Scott R.J., *Rec. Trav. Chim.*, 1956, vol. 75, pp. 787 - 789.

[92] Haldna U. and Mursxhan A., Izv. AN EstSSR, vol. 32(I), pp. 47 - 49. (Rus)

[93] Luk'yanets V.M., Makitra R.G., and Tsikanchuk Ya.M., *ZhOKh*, 1977, vol. 47(3), pp. 506 - 513. (Rus)

[94] Luk'yanets V.M., Makitra R.G., and Tsikanchuk Ya.M., *ZhOKh*, 1977, vol. 47(5), pp. 1084 - 1085. (Rus)

[95] Makitra R.G. and Pirig Ya.N., *Reakts. Spos. Org. Soed.*, 1979, vol. 16(4), pp. 497 – 507. (Rus)

[96] Frenkel Ya.I., 'Kinetic theory of liquids', In: *Selectas*, vol. 3, Moscow-Leningrad, Izd. AN SSSR, 1959, p. 181. (Rus)

[97] Paetzold R. and Niendorf K., *Z. Anorg. Allg. Chemie*, 1974, Bd, 405(2), S. 129 - 131.

[98] West R., Powell D.L., Whatley L.S., Lee M.K., and von Schleyer P., *J. Am. Chem. Soc.*, 1962, vol. 84(16), pp. 3221 - 3223.

[99] Joesten M.D. and Drago R.S., *J. Am. Chem. Soc.*, 1962, vol. 84(20), pp. 3817 - 3821.

[100] Epley T.D. and Drago R.S., *JAC 5*, 1967, vol. 89(21), pp. 5770 - 5773.

[101] Trotter P.J. and Hanna M.W., *J. Amer. Chem. Soc.*, 1966, vol. 88(16), pp. 3724 - 3729.

[102] Baranovskaya O.E., Makitra R.G., and Pirig Ya.N., Reakts. Spos. Org. Soed., 1977, vol. 14(1), pp. 51 - 66. (Rus)

[103] Dyankov S.S. and Iliew V.I., *Rep. Bulg. Acad. Sci.*, 1971, vol. 24(5), pp. 605 - 608.

[104] Dyankov S.S. and Iliew V.I., *Rep. Bulg. Acad. Sci.*, 1971, vol. 24(7), pp. 885 - 887.

[105] Dyankov S.S. and Iliew V.I., *Rep. Bulg. Acad. Sci.*, 1971, vol. 24(1), pp. 1051 - 1054.

[106] Iliew V.I. and Dyankov S.S., *Rep. Bulg. Acad. Sci.*, 1976, vol. 29(3), pp. 376 - 378.

[107] Mouchet I.P. and Rousset J., *J. Chim. Phys. et Phys. Chim. Biol.*, 1980, vol. 77(6), pp. 529 - 530.

[108] Nagy O.B., Nagy I.B., and Bruylants A., *J. Chem. Soc.*, 1972, Perkin Trans., part 2, No. 8, pp. 968 - 972.

[109] Nagy O.B., Nagy I.B., and Bruylants A., *Bull. Chem. Soc. Belge*, 1974, vol. 83(3/4), pp. 163 - 171.

[110] Dupire S., Mulindalynama J.M., Nagy J.B., and Nagy O.B., *Tetrahedron*, 1975, vol. 31(4), pp. 767 - 773.

[111] Makitra R.G. and Pirig Ya.N., *Ukr. Khim. Zh.*, 1986, vol. 52(9), pp. 907 - 911. (Rus)
[112] Walkow P. and Epperlen J., *J. Prak. Chem.*, 1984, Bd. 326(1), S. 63 - 72.
[113] Makitra R.G. and Pirig Ya.N., *Ukr. Khim. Zh.*, 1988, vol. 54(7), pp. 747 - 751. (Rus)
[114] Pytela O., *Collect. Czech. Chem. Commun.*, 1988, vol. 53(7), pp. 1333 - 1423.
[115] Searles S. and Tamres M., *J. Am. Chem. Soc.*, 1951, vol. 73(8), pp. 3704 - 3706.
[116] Tamres M., Searles S., and Leighly E.M., *J. Am. Chem. Soc.*, 1954, vol. 76(15), pp. 3983 - 3985.
[117] Mel'nik A.P., Passet B.V., Alobeinikov A.A., and Gievoi G.M., *Reakts. Spos. Org. Soed.*, 1973, vol. 10(3), pp. 665 – 674. (Rus)
[118] Krestov G.A., *Voprosy Fiz. Khim.*, Iss. 10, Vestnik Kharkovskogo Univ., 1979, No. 192, pp. 12 - 21. (Rus)
[119] Krestov G.A., Zakharov L.G., and Romanov V.A., *Izv. VUZov, Khim. I Khim. Tekhnol.*, 1976, vol. 19(9), pp. 1276 - 1277. (Rus)
[120] Makitra R.G. and Pirig Ya.N., *Reakts. Spos. Org. Soed.*, 1979, vol. 16(1), pp. 103 – 107. (Rus)
[121] Tsvetkov V.G., Kupriyanov V.F., Novoselova I.V., and Klimin V.V., *ZhOKh*, 1977, vol. 47(4), pp. 729 – 731. (Rus)
[122] Tsvetkov V.G., *ZhOKh*, 1980, vol. 50(2), pp. 258 – 262. (Rus)
[123] Makitra R.G., Pirig Ya.N., and Tsvetkov V.G., *ZhOKh*, 1983, vol. 53(4), pp. 727 – 732. (Rus)
[124] Shilov E.A., Serguchev Yu.A., Sergeev G.B., Smirnov V.V., and Kofman V., *Doklady AN SSSR*, 1971, vol. 197(6), pp. 1096 - 1099. (Rus)
[125] Drago R.S. and Wenz D.A., *J. Amer. Chem. Soc.*, 1962, vol. 84(4), pp. 526 - 528.
[126] Maes G. and Zeegers-Huyskens.T., *J. Raman Spectr.*, 1978, vol. 7(6), pp. 325 - 329.
[127] Makitra R.G. and Pirig Ya.N., *Reakts. Spos. Org. Soed.*, 1979, vol. 16(4), pp. 512 – 520. (Rus)
[128] Makitra R.G., Pirig Ya.N., and Tsvetkov V.G., *ZhOKh*, 1984, vol. 54(7), pp. 1449 – 1455. (Rus)
[129] Makitra R.G., Pirig Ya.N., and Tsvetkov V.G., *ZhOKh*, 1984, vol. 54(4), pp. 833 – 836. (Rus)
[130] Tsvetkov V.G., Buslaeva M.N., and Kruglyak L.I., *Reakts. Spos. Org. Soed.*, 1980, vol. 17(2), pp. 129 – 136. (Rus)
[131] Naberukhin Yu.I., Buslaeva M.N., and Dudnikova K.T., *Koord. Khim.*, 1982, vol. 8(1), pp. 29 - 30. (Rus)
[132] Makitra R.G. and Pirig Ya.N., *Ukr. Khim. Zh.*, 1980, vol. 46(1), pp. 83 - 88. (Rus)
[133] Tamres M., *J. Amer. Chem. Soc.*, 1952, vol. 74(13), pp. 3575 - 3578.
[134] Pimentel J. And Mac-Clellan O., *Hydrogen Bond*, Moscow, Mir, 1964, pp. 77 – 88. (Rus)
[135] Makitra R.G. and Pirig Ya.N., *Reakts. Spos. Org. Soed.*, 1979, vol. 16(2), pp. 159 – 164. (Rus)
[136] Makitra R.G., Pirig Ya.N., and Tsvetkov V.G., *ZhOKh*, 1986, vol. 56(7), pp. 1452 – 1457. (Rus)
[137] Stephenson W. and Fuchs R., *Canad. J. Chem.*, 1985, vol. 63(2), pp. 336 - 342.
[138] Stephenson W. and Fuchs R., *Canad. J. Chem.*, 1985, vol. 63(2), pp. 342 - 348.
[139] Makitra R.G. and Pirig Ya.N., *Zh. Prikl. Spektr.*, 1984, vol. 41(4), p. 690. (Rus)

In: Frontiers in Physical Organic Chemistry
Editor: G.E. Zaikov, pp. 157-194

ISBN 1-60021-128-3

Chapter 8

KINETIC PROBING OF NON-CRYSTALLINE POLYMERIC MATRIX GLASSY-LIKE POLYMERS WITH THE HELP OF DIBENZOYL PEROXIDE

Yu. A. Mikheev and G. E. Zaikov
N.M.Emanuel Institute of Biochemical Physics Russian Academy of Sciences,
4, Kosygin st., Moscow 119991, Russia

1. JUSTIFICATION OF THE PROBE REACTION SELECTION

The supramolecular model of the polymeric microreactor, developed on the basis of the broad experimental material, allowed formulation of the key structural-kinetic stage of the chain reaction [1 - 7]. In accordance with this model, all quasi-elementary stages of any chain reaction, proceeding in non-crystalline polymeric matrix (NCPM) microreactor, are display highly heterogeneous physical origin. This drastically distinguishes the non-crystalline matrix of polymers from homogeneous heterophase and liquid-phase systems.

In gases and liquids, chemical activity of particles primarily depends upon structural and energy properties of activated transient complex formed from the reacting particles. The reaction rates in these phases are determined by the probabilities of activated complex formation and overcoming the energy barrier by atoms of the complex during translation along the reaction coordinate [1]. In contrast with gases and liquids, chemical process in polymers proceeds under qualitatively different conditions, associated with the specific structural organization of polymeric chains shaped as nanoreactors. Being typical elements of polymeric nanoreactors (*gs*-micelles) and possessing specific force fields, the nanoreactors modulate chemical activity of reacting particles versus the structural-physical state of *gs*-micelles. Hence, the character of elementary chemical act, proceeding in nanoreactors-nanopores, is qualitatively varied in relation to homogeneous reaction acts: this act becomes heterogeneous, because reacting particles in it are physically connected to the nanopore and its chain-fringed environment, which is absent for gases and liquids.

For NCPM molecular-chain sponge, a situation is typical, when identical reaction centers are able to react in different-sized nanopores under the effect of unequal intermolecular forces. In this case, the factor of structural-physical reactivity modulation can be expressed not only quantitatively (by occurrence of aggregates of particles with unequal reaction rate constants, i.e. in the form of the so-called polychronic reactivity), but also qualitatively, when in differently sized nanoreactors from the same active particles different reaction products are produced.

In relation to logics of the structural-kinetic model [7 - 9] the initiated polymeric chain reaction can be defined as a heterogeneous-heteronanophase reaction which kinetic units are distributed by inhomogeneous nanoreactors (or structural zones with inhomogeneous nanoreactors).

In the early works, experimental checking of the above-described kinetic model (hereinafter, for brevity sake it will be called the heterophase one) was performed by the method of the probe chain reaction. Generally speaking, any reaction may be taken as a probe one, the rate of which varies highly with the change of polymeric *gs*-micelle physical state. However, the highest information value (and validity at the initial stage of structural-kinetic model development) must be displayed by reactions, in which not only the kinetic index (the rate constant), but also qualitative parameter, i.e. the kinetic equation of the reaction, changes at the transition from liquid solutions to polymers.

In this connection, it is desirable to choose both polymeric and light free radicals as carriers of probe reaction chains; moreover, they both must react with macromolecules and initiator molecules. In this case, potential possibilities will be provided in the reacting system which, on the one hand, realize qualitatively different translation processes for free valences and, on the other hand, competitive reactions of free radicals differing by molecular mass and migration methods by the matrix of *gs*-micelles.

In accordance with the above-said, at the initial stages of kinetic model development chain reactions with dibenzoyl peroxide was preferred. Dibenzoyl peroxide (BP) is one of the most frequently used sources of free radicals in kinetic studies. This reagent is not only the initiator of various chain reactions, but also the acceptor of free radicals. This is the reason why it itself is capable of decaying by the chain mechanism in the case, if dissolved oxygen is absent in the system. The broad data on kinetics of BP chain degradation in homogeneous liquids are present in the literature that stipulates the final choice made for the benefit of application of the current model reaction for kinetic probing of polymers.

The main objective of the work consisted in studying the way of the kinetic law (and product content) variation for anaerobic dibenzoyl peroxide degradation at homogeneous liquid replacement by a polymer, which plays the role of high-molecular homologue of the liquid. The present Chapter discusses heterophase mechanisms of the reactions studied using glassy-like samples of polycarbonate (PC), cellulose triacetate (CTA), polystyrene (PS), glassy-like and rubbery polyamide PA-548, rubbery atactic polypropylene (APP) and polyisobutylene (PIB), polyethylene oxide (PEO) melt, as semicrystalline polyethylene (PE) and polypropylene (PP).

The occurrence of BP chain reactions in polymers (in the presence of oxygen) was first observed for PS, PE and PP with the help of non-kinetic dose technique using the dependence of decomposed BP dose on the initial BP content and that for the accumulated reaction products (PhCOOH, CO_2) [2 - 5]. The peculiarities found stimulated the authors [2 - 5] to postulate the second order of the chain BP reaction by BP and its induction by the attack of free benzoyloxyl on BP, producing phenylbenzoate.

However, the following detailed study of the kinetics and composition of anaerobic BP reaction products performed with various polymers showed induction of reaction chains by polymeric free radicals for all cases, phenylbenzoate output in the reaction chains being negligibly low. Most frequently, the reaction in polymers is of pseudo-first order by BP and its effective rate constant depends on the initial BP concentration. In all cases, the kinetic law of the reaction in NCPM is changed qualitatively in relation to that of the liquid-phase BP reaction, the reaction product composition changes, the frequency of free benzoyloxyls' decarboxylation acts decreases, and besides the property of terminating kinetic chains, accumulated low-molecular phenyl compounds become capable of promoting BP chain decomposition, which is untypical of liquids. All non-liquid kinetic features of BP reaction with polymers have been explained in the framework of the heterophase reaction model. The mechanism of occurrence of qualitatively new kinetic features in this reaction as liquid is replaced by a polymer will be discussed below in this Chapter.

The real effect of the structural-physical state of polymeric matrix on BP reaction kinetics is clearly observed in experiments needless to compare it with the liquid-phase process kinetics. For example, the effective rate constant of BP decomposition in CTA samples varies in a broad range with thermodynamic quality of the solvent [6], used for sample preparation. The samples [6] were prepared as films using:

i) pure thoroughly dried methylene chloride;
ii) water-saturated methylene chloride;

as well as a fibrous wool, obtained by precipitation of the polymer with BP from solution in methylene chloride by 1:1.5 water - methanol mixture.

Mechanical stretch of PC films containing BP in a flow of pure carbon dioxide increases BP decomposition rate and the rate of macromolecule breaks since the very beginning of creep deformation [7]. Depending on temperature and applied stress, the initial BP decomposition rate may increase by 7 – 15 times, whereas the rate of macromolecule breaks is sensitive to mechanical load. It has been shown [7] that in the case of mechanically induced structural relaxation in polycarbonate films, BP reaction cannot be kinetically described in the framework of simple reaction layouts. Meanwhile, high sensitivity of the process itself to structural-mechanical impact became an additional stimulus for using this reaction for kinetic probing of polymers in experiments, where the stationary state of NCPM structure is preserved.

Glassy-like polymeric samples designed for kinetic probing represented films (20 - 50 μm thick), prepared from solutions in dry methylene chloride. Under BP concentration used the films preserved high optical transparency of the initial polymers in the visible spectrum region, because added BP did not solidify. The specimens were heated in sealed evacuated glass tubes as well in tubes with air. The current BP contents were determined using the iodimetric method of spectrophotometry. In some cases, decomposition products were analyzed using the fine-layer chromatography. The number of polymer chain scissions was calculated by the data of viscosimetry of polymer solutions in methylene chloride.

2. Homogeneous Mechanism for Dibenzoyl Peroxide Reaction With Liquids

BP reacts with liquid hydrocarbons by the free radical mechanism, and the reaction begins with dissociation into benzoyloxyl radicals:

$$\text{PhCOO-OOCPh} \xrightarrow{k_0} 2\text{PhCOO}^\bullet\ (\text{r}^\bullet). \qquad (2.0)$$

Further destiny of benzoyloxyl radicals depends on the ratio of their decarboxylation rates:

$$\text{r}^\bullet \xrightarrow{k_1} \text{Ph}^\bullet + CO_2 \qquad (2.I)$$

and hydrogen atom detachment from alkyl fragments of the solvent molecules:

$$\text{r}^\bullet + \text{RH} \xrightarrow{k_2} \text{rH} + \text{R}^\bullet. \qquad (2.II)$$

Alkyl radical $R^\bullet$, which is also eliminated from the solvent by free phenyl radical as follows:

$$\text{Ph}^\bullet + \text{RH} \xrightarrow{k_2^*} C_6H_6 + \text{R}^\bullet, \qquad (2.II^*)$$

attacks the benzene ring in dibenzoyl peroxide, thus initiating BP decomposition:

$$\text{R}^\bullet + \text{BP} \xrightarrow{k_3} \text{r}^\bullet + \text{products}, \qquad (2.III)$$

and recombines colliding with another radical.

The total BP decomposition rate is summed up from the rates of spontaneous and chain reactions [8]:

$$-\frac{dC}{dt} = k_0 C + k_{\text{ch}} C^n,$$

where C is the current BP concentration; k_0 and k_{ch} are the rate constants for spontaneous and induced channels of BP decomposition; n is the kinetic order of the chain reaction.

The specific shape of coefficients k_{ch} and n of the chain reaction term depends on the propagation steps (induced BP decomposition), in which chain termination reactions dominate in the given liquid. In simple kinetic schemes proposed in [8], radicals of the same type ($r^\bullet$ or $R^\bullet$) participate in both propagation and termination of the reaction chains. In the case of alkyl radicals $R^\bullet$ dominance, it is enough to take a combination of reactions (2.III) and

$$R^{\bullet} + R^{\bullet} \xrightarrow{k_4} \text{(termination)}, \tag{2.IV}$$

and theoretically determined BP decomposition rate is reduced to the following form:

$$-\frac{dC}{dt} = k_0 C + k_3 \left(\frac{k_0}{k_4}\right)^{0.5} C^{1.5}. \tag{2.1}$$

The appropriate expressions for other possible combinations of the reactions were deduced in [9, 10]. In particular, if the reaction (2.III) is attended with termination of radicals, ($r^{\bullet}$ + $r^{\bullet}$) or ($r^{\bullet}$ + $R^{\bullet}$), the reaction order n by peroxide equals 0.5 or 1.0, respectively. On the assumption that BP decomposition is initiated by benzoyloxyls, n becomes equal 1.5, 2.0 and 2.0, depending on the dominant channel of radical termination ($r^{\bullet}$ + $r^{\bullet}$), ($r^{\bullet}$+$R^{\bullet}$), or ($R^{\bullet}$ + $R^{\bullet}$).

For liquids, the most frequent order of the chain reaction is n = 1.5, as in the equation (2.1) [8]. The order 1 is infrequently observed, while values 0.5 and 2.0 are probably unknown. This may be caused by high rates of (2.I), (2.II), and (2.II*) reactions providing the inevitability of stages (2.III) and (2.IV). Actually, the decarboxylation rate constant k_1 is fairly high: 1×10^8 s^{-1} at 403 K and 6×10^5 s^{-1} at 313 K; the activation energy equals E_1 = 58.6 kJ/mol [11]. The reaction (2.II) involving secondary CH group of hydrocarbons displays the activation energy E_2 = 36.4 kJ/mol and may occur in parallel to the reaction (2.I) [12]. Considering high rates of reactions (2.I) and (2.II), the attack of $r^{\bullet}$ radicals at benzene rings in BP dissolved in aliphatic hydrocarbons must be inefficient.

As follows from the experimental results on benzene solutions, free benzoyloxyl radicals interact with C_6H_6 molecules forming a σ-complex which rapidly and reversibly dissociates to the initial reagents – benzene and radical $r^{\bullet}$ [13, 14]. At 353 K, equilibrium in this reversible reaction was considerably shifted toward dissociation. Falling temperature enhances the role of other reactions with the σ-complex (r - $C_6H_6^{\bullet}$), e.g. of its oxidation by dissolved oxygen to produce tangible quantities of phenylbenzoate [13, 14]. Moreover, owing to the reversibility of benzoyloxyl addition to benzene; $r^{\bullet}$ in benzene solution has a good chance to react with BP forming the r – $BP^{\bullet}$ complex which reversibly dissociates and partly converts into disalicylic ester [15]. This reaction pathway occurs, when oxidizing r – $BP^{\bullet}$ to r – BP (r – $C_6H_4COOOOCPh$), and then dissociation of r – BP followed by cyclization of r – $C_6H_4COO^{\bullet}$ to σ-complex by attacking the attached phenyl ring of r. Thereafter, the σ-complex is oxidized by BP to disalicylic ester.

So far as concerns phenyl radicals, they easily detach hydrogen atoms from aliphatic hydrocarbons, the rate constant k_2* = (0.2 - 6.2)$\times10^6$ $M^{-1}s^{-1}$ being significantly higher than for similar reaction with alkyl radicals (10 - 10^3 $M^{-1}s^{-1}$) [16].

An attack of free phenyl radicals on dissolved BP yields in a minor amount of phenylbenzoic acid, because of their much faster interact with a solvent. Nonetheless, their reaction with BP should sometimes be taken into account, e.g. to account for the isotopic composition of the reaction products of labeled compounds [17].

The reaction mechanism (2.III) depends on the chemical origin of $R^{\bullet}$ radical. For instance, free ether radicals preferably attack oxygen of the peroxidic bond [18, 19]:

$ROC^{\bullet}HR_1 + BP \rightarrow ROCH(R_1)OCOPh + r^{\bullet}.$

If the free valence on $R^{\bullet}$ is not adjacent to O atom, BP phenyl ring is favorably attacked [20 - 22]:

$R^{\bullet} + BP \rightarrow RC^{\bullet}{}_6H_5COOr$ (σ-complex) $\rightarrow RC_6H_4COOH + r^{\bullet}$;

$RPh + CO_2 + r^{\bullet}.$

The nascent σ-complex subsequently either undergoes disproportionation by reacting with another radical or experiences an intramolecular transfer of hydrogen atom and peroxide bond break.

Almost all benzoyloxyl radicals arisen by BP decomposition in a benzene solution undergo decarboxylation, and the nascent free phenyl radicals predominantly add themselves to benzene forming σ-complexes:

$Ph^{\bullet} + C_6H_6 \rightarrow PhPhH^{\bullet}.$

This complex attacks BP, undergoing oxidation with a hydrogen atom transfer

$PhPhH^{\bullet} + BP \rightarrow PhPh + rH + r^{\bullet}$

and participating in reaction chain termination. The expression for the peroxide decay rate is somewhat more complex than in the case of nonaromatic hydrocarbons:

$$-\frac{dC}{dt} = k_0C + k_{iL}C + k_iC^{1.5}.$$

It indicates the occurrence of two chain reactions of the order 1 and 1.5 in BP. However, the first chain reaction is inefficient, for its rate constant ($k_{iL} = 0.2\times10^{-5}$ s^{-1} at 351 K) is well below the rate constant for spontaneous decomposition ($k_0 = 1.67\times10^{-5}$ s^{-1}) [23].

BP chain decomposition in alkylbenzene solutions is induced by σ-complexes of phenyl and the solvent, $PhC_6H_5^{\bullet}$-Alk, and also by arylalkyl radicals C_6H_5-Alk$^{\bullet}$, and its rate is summed up from those of spontaneous reaction and the chain reaction of the order 1.5 by BP [24]. As opposed to alkylbenzenes, chlorobenzene and bromobenzene are infrequent examples of solvents, where the chain reaction is intense and the order one by BP, which is attributable to cross termination of the reaction chains occurring in these media, as radicals $r^{\bullet}$ encounter σ-complexes of halogenated benzene [25].

The rate of BP chain decomposition is the lowest in benzene, toluene, and ethylbenzene and increases as transiting to non-aromatic solvents. The ratios of the initial rates of chain and spontaneous reactions,

$$\frac{W_i^0}{W_0^0} = \frac{k_iC_0^{0.5}}{k_0} \quad (353\text{ K}, C_0 = 0.2\text{ M}),$$

estimated in [8], offered a possibility of arranging these solvents in a series according to their efficiency: benzene - 0.58, toluene - 0.58, carbon tetrachloride - 0.79, cyclohexene - 1.07, ethyldiiodide - 0.75, tert-butylbenzene - 2.09, cyclohexane - 2.29, ethylacetate - 1.82, acetic acid - 2.81, and acetanhydride - 1.82.

Low rates of the chain decomposition in benzene and toluene are associated with a high probability of quadratic reaction chain termination on encounter of two σ-complexes. Because of high reactivity of phenyl-containing compounds in radical termination reactions, as built-up, the BP decomposition products in non-aromatic liquids retard the chain process. In acetanhydride, e.g. the chain reaction efficiency reduces from its initial value of 1.82 to 0.59 at high BP decomposition degrees [8].

The spontaneous reaction contribution into BP decomposition process increases with peroxide decomposition temperature, because the activation energy Ei for the chain reaction is less than E0. For example, in benzene Ei = 105.5 kJ/mol, and in acetanhydride 104.5 kJ/mol [8]. On the other hand, the available values of preexponential factor and activation energy of the rate constant for the spontaneous reaction,

$$k_0 = A \exp\left(-\frac{E_0}{RT}\right),$$

are: A = 1.18×1014 s–1, E0 = 125 kJ/mol [10] and A = 3×1013 s–1, E0 = 124 kJ/mol [26], respectively (higher E0 = 138 - 139 kJ/mol were observed in [8, 27]). Thus, the efficiency of the chain component in the liquid-phase process is temperature dependent, as observed from the theoretical expression for its rate. According to it, the composite activation energy Ei equals a half of E0 only.

So far as concerns the stoichiometric coefficient of free radical generation, it is close to its maximum in slightly viscous liquids: benzene, e.g. every event of spontaneous BP decomposition produces two free radicals. This follows from the data of independent studies at 351 K [23] and 318 K [28].

Among usual solvents, ethers and alcohols perform the fastest react with BP. In alcohols, the induced BP decomposition is associated with the transfer of hydrogen atom from alcoholic radical to peroxide group [10, 29]:

$$RCH_2O^{\bullet} \leftrightarrow RC^{\bullet}HOH;$$

$$RC^{\bullet}HOH + BP \rightarrow RCHO + PhCOOH + PhCOO^{\bullet}\ (r^{\bullet}).$$

The attachment of free alkoxy-radical to phenyl ring in BP is also possible. This is evidenced synthesis of both aldehyde and a great amount of isobutoxybenzoic acids in BP interaction with isobutanol [30].

Completing the review of the liquid-phase process, it is desirable to note high chain reaction sensitivity to oxygen dissolved in liquids. Oxygen effectively inhibits the chain process by converting active radicals into peroxy ones which do not react with dibenzoyl peroxide [8, 31].

3. Heterophase Mechanism of Cellulose Triacetate and Polycarbonate Arylation by Benzoyl Peroxide

3.1. Kinetic Phenomenology of the Thermal Reaction [32, 33]

Heating of BP doped CTA and PC specimens in sealed evacuated glass tubes induces the BP decomposition in accordance with the exponential law as follows:

$$C = C_0 \exp(-kt), \qquad (3.1)$$

where C_0, C are the initial and the current BP concentrations, respectively; k is the effective rate constant of BP decomposition.

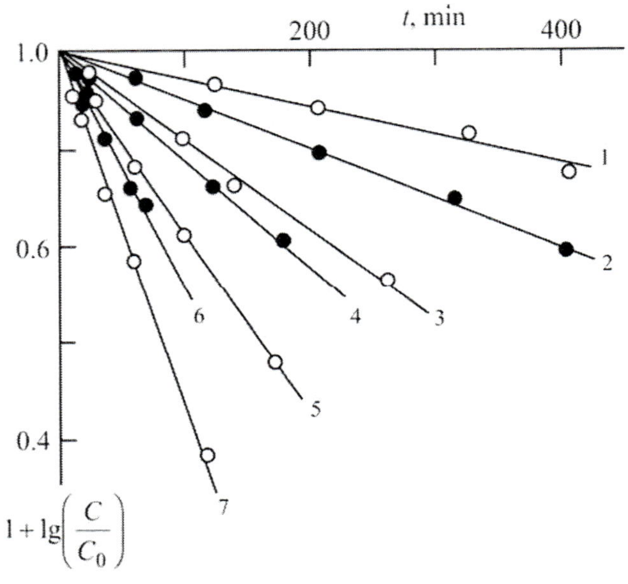

Figure 1. Kinetic curves of BP decomposition in evacuated CTA films plotted in semi-logarithmic oordinates for temperatures (1) 353, (2) 359, (3) 365, (4)-(7) 371 K; initial BP concentrations, C_0: (4) 0.06, (5) 0.196, (1) - (3), (6) 0.38, and (7) 0.69 mol/kg

Direct anamorphoses of BP decomposition curves plotted in coordinates of the equation below:

$$\lg\left(\frac{C}{C_0}\right) = 1 - 0.43kt,$$

are shown in Figure 1 (for CTA) and Figure 2 (for PC) at temperature 371 K.

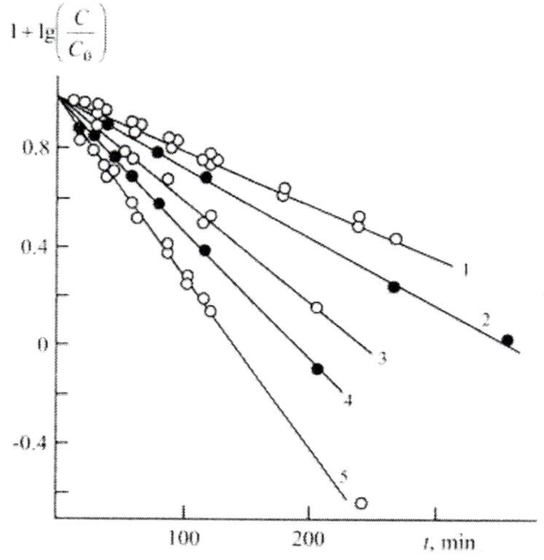

Figure 2. Kinetic BP decomposition curves ($T = 371$ K) plotted in semi-logarithmic coordinates for (1) aerated (by ambient air) and (2) - (5) evacuated PC films containing BP at initial concentration, C_0: (1) 0.12, 0.21, 0.41, 0.7, 0.83; (2) 0.1; (3) 0.33; (4) 0.5; (5) 0.83 mol/kg

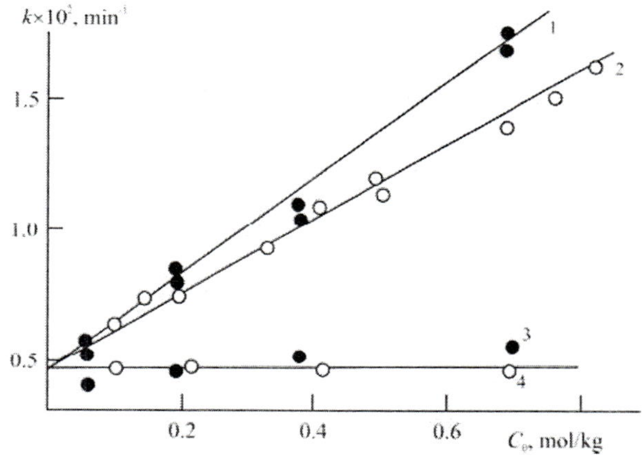

Figure 3. Dependence of constant k on C_0 at 371 K in evacuated (1, 2) and aerated (3, 4) CTA films (1, 3) and PC films (2, 4)

For both polymers, constant k is linearly dependent on C_0 (Figure 3, curves 1 and 2):

$$k = k_0 + k_0 k_i C_0, \tag{3.2}$$

where k_0 is the rate constant of nonchain BP decomposition, which is calculated from the cutoff on the axis of ordinates in Figure 3; k_i is the chain reaction constant. The same k_0 value characterizes the exponential curves of BP decomposition in the films aerated by ambient air (Figure 3, curves 3 and 4). As lines 1, 2 and 3, 4 (Figure 3) are compared, one observes that oxygen inhibits BP chain reaction at eliminating the effective chain rate constant k_0k_i and producing no polymer peroxides being titrated.

In the absence of O_2, BP decomposition rates in nonaromatic CTA and aromatic PC are accepted to be nearly equal. In both cases at 371 K, constants in equation (3.2) are the following: $k_0 = (4.7 \pm 0.5)\times10^{-3}$ min^{-1}, the same for CTA and PC; $k_0k_i = (1.8 \pm 0.2)\times10^{-}$

^{2} kg/mol·min (CTA) and 1.4×10^{-2} min^{-1} (PC). At 353 K in PC [34], $k_0 = (7 \pm 1)\times10^{-4}$ min^{-1} and $k_0k_i = (10 \pm 1)\times10^{-4}$ kg/mol·min. The effective activation energy of nonchain BP decomposition is $E_0 = 117$ kJ/mol, the same for CTA [35] and PC that is somewhat lower in the liquid phase.

The effective activation energy of the chain reaction $E_i = 160$ kJ/mol is higher than E_0 (whereas in liquids the opposite is true: $E_i < E_0$; refer to Section 2).

BP decomposition induces degradation of both polymers pursuant to the order one kinetic law by BP. Thus, the chain break rate is:

$$\frac{dN}{dt} = k_N C = k_N C_0 \exp(-kt), \quad (3.3)$$

and the increase in the amount of polymer chain breaks follows the formula as follows:

$$N = \frac{k_N C_0}{k}[1 - \exp(-kt)] = N_\infty[1 - \exp(-kt)]. \quad (3.4)$$

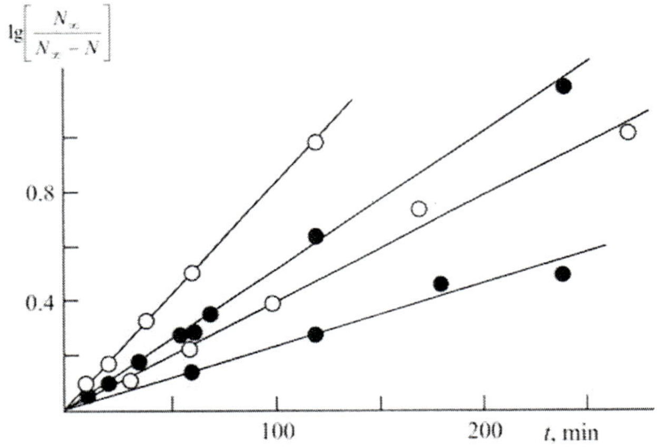

Figure 4. Kinetic plots of polymer backbone scissions induced by BP in evacuated CTA films at $T = 371$ K; . initial BP concentration, C_0: (1) 0.06, (2) 0.196, (3) 0.38, and (4) 0.69 mol/kg

Figure 4 shows the plots of equation (3.4) reconstructed for the straightforward form:

$$\lg[N_\infty(N_\infty - N)] = kt.$$

The rate constant k values, calculated by these plots, correspond to those found from BP decomposition kinetics. Here N is the number of broken chemical bonds per initial CTA macromolecule with $M_n = 330{,}000$. The constant k_N, calculated from the limiting number of breaks N_∞ at 371 K, equals 0.1 kg·mol^{-1}min^{-1}macromolecule^{-1} that gives $k_n = 3.0\times10^{-4}$ min^{-1} after recalculation per molar concentration of the chain breaks (n, mol/kg).

Kinetic curves of polymer chain break for PC are shown in Figure 5 (371 K). They are satisfactorily described by the first-order equation by BP [34]:

$$n = n_\infty[1 - \exp(-kt)], \quad (3.5)$$

where $n_\infty = \dfrac{k_n C_0}{k}$ and $k_n = 6.9\times10^{-4}$ min^{-1}. The constants k and k_n are independent of PC molecular mass (experimental M_{vo} values [34] equal 128,000, 113,000, 100,000, 64,000, and 35,500).

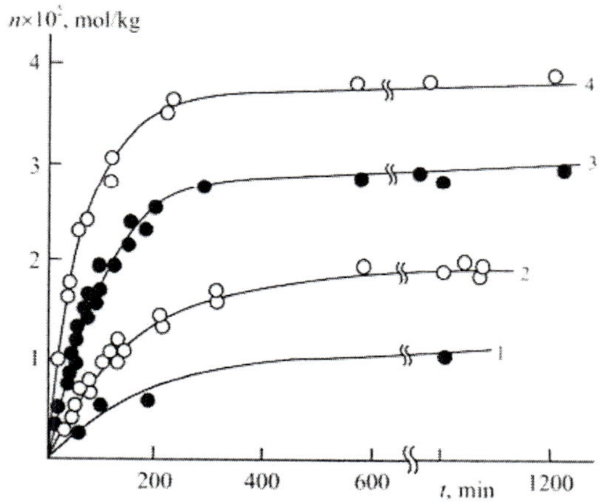

Figure 5. Kinetic curves of polymer main chain scissions induced by BP in evacuated PC films at $T = 371$ K; initial BP concentration C_0: (1) 0.083, (2) 0.19, (3) 0.45, and (4) 0.83 mol/kg

In films aerated by ambient air, the polymer chain break rate and the kinetic curves are described by the following expressions:

$$\frac{d(n^a)}{dt} = k_n^a C = k_n^a C_0 \exp(-k_0 t),$$

$$n^a = n_\infty^a [1 - \exp(-k_0 t)], \qquad (3.6)$$

where $n_\infty^a = \dfrac{k_n^a C_0}{k_0}$.

Figure 6 shows the linear plot for equation (3.6), which gives a good opportunity to calculate the effective constant $k_n^a = (3 \pm 1)\times10^{-4}$ min^{-1}.

The constant k_n^a for aerated CTA films, calculated by the data published elsewhere [35], is $(1.3 \pm 0.3)\times10^{-4}$ min^{-1}. For both polymers, there is a correlation:

$$k_n^a = 0.5 k_n.$$

During BP decomposition, benzoic acid (rH), as well as arylated macromolecules and a small amount of phenyl benzoate (rPh) are accumulated in the evacuated polymer films [32]. It has been assumed [32] that the final product X is formed both in nonchain and chain reactions with probabilities α_x and β_x, respectively. In such a case, the accumulation rate of X equals:

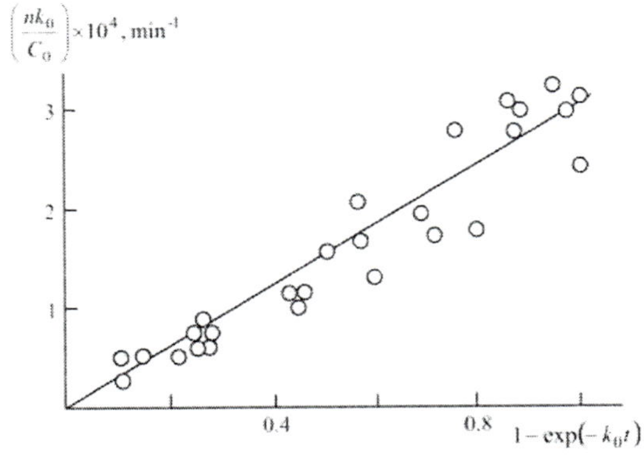

Figure 6. Kinetic curves of PC backbone scissions induced by BP in aerated (by ambient air) films at T = 371 K and initial BP concentration C_0 of 0.13 - 0.65 mol/kg; polymer M_v = 25,000 – 128,000

$$\frac{d[X]}{dt} = \alpha_x k_0 C + \beta_x k_0 k_i C_0 C. \tag{3.7}$$

Integrating (3.7) with respect to (3.1), one can deduce an expression for the current concentration as follows:

$$[X] = \frac{\alpha_k k_0 + \beta_x k_0 k_i C_0}{k} C_0 [1-\exp(-kt)],$$

and for the ultimate product concentration (at $t \to \infty$):

$$[X]_\infty = \frac{\alpha_k k_0 + \beta_x k_0 k_i C_0}{k} C_0. \tag{3.8}$$

The expression (3.8) is reduced to

$$\frac{k[X]}{C_0} = \alpha_k k_0 + \beta_x k_0 k_i C_0, \tag{3.9}$$

for the purpose of determining α_x and β_x coefficients from the straight lines plotted. The plots of equation (3.9) are shown in Figure 7. Using these plots, the reaction synthesizing X product can be simply determined. Clearly in particular, CTA and PC degradation is only associated with the non-chain decay of BP: $\beta_n = 0$ and $k_n = \alpha_n k_0 \approx 3\times10^{-4}$ min^{-1} (for CTA); $k_n = \alpha_n k_0 \approx 6.9\times10^{-4}$ min^{-1} (for PC).

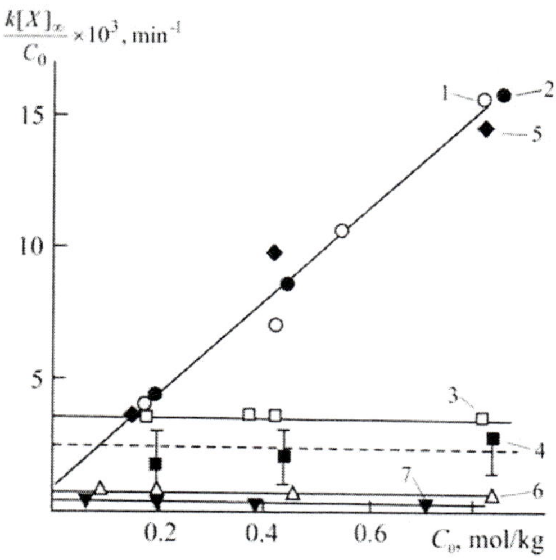

Figure 7. Dependence of normalized yield of BP reaction products on initial BP concentration C_0 for (1), (3), (5), (7) CTA and (2), (4), (6) PC films: (1), (2) benzoic acid; (3), (4) phenylbenzoate; (5) CTAA (arylated polymer); (6), (7) backbone scission for PC and CTA, respectively

The rH product is synthesized from CTA and PC in almost equal amounts by both non-chain ($\alpha_{rH}k_0 = 1\times10^{-3}$ min^{-1}) and chain reactions ($\beta_{rH}k_0k_i = 2\times10^{-2}$ kg·mol^{-1}min^{-1}). The output of rH in the non-chain reaction is low ($\alpha_{rH} = 0.2$), reaching its maximum in the chain reaction ($\beta_{rH} = 1$).

Phenyl benzoate is mainly produced by the non-chain reaction, where for CTA $\alpha_{rPh} = 0.8$ and $\beta_{rPh} = 0$ and for PC $\alpha_{rPh} \approx 0.5$ and $\beta_{rPh} \approx 0$. The condition that for PC $\beta_{rPh} \approx 0$ is substantiated by equal values of benzoic acid output, maximal for both CTA and PC ($\beta_{rH} = 1$).

The portion of low molecular weight products of the non-chain BP decomposition corresponds to ($\alpha_{rH} + \alpha_{rPh}) \approx 1$ for CTA and ~0.7 for PC. Probably, the benzene rings of PC capture parts of free radicals synthesized during non-chain BP decomposition.

High transparency of CTA in UV region of the spectrum with $\lambda > 260$ nm allows recording of UV absorption of the CTA arylation product (CTAA) [32, 33] and obtaining of the values $\alpha_{CTAA} \to 0$, $\beta = 1$. For CTAA, relative to alkylbenzenes and alkylbenzoic acids the molar extinction coefficient equals (1,100 ± 200) l·mol^{-1}cm^{-1}. Figure 7 also shows the experimental data for CTAA.

3.2. Peculiarities of Polymer Reaction With BP

In evacuated glassy CTA and PC films, the chain BP reaction can get the same efficiency, as for carboxyl containing liquids

$$\frac{W_i}{W_0} = \frac{k_0 k_i C_0}{k_0} = 2 \div 2.5.$$

Clearly the extreme dominancy of polymers against liquids with respect to macroscopic viscosity (over 10^{14} times) makes no obstacles for bimolecular acts of induced BP decomposition. In this respect, the polymers are identical liquids, providing high mobility for low molecular weight particles, such as BP and benzoyloxyl.

The additives, however, do not diffuse from the evacuated films. They disappear exclusively in the chemical process, which is proved by the material balance $\beta_{BP} = \beta_{rH} = \beta_{CTAA} = 1$. (In terms of decomposed BP material balance, i.e. $[X]_\infty = C_0$, equations (3.2) and (3.9) give the relations $\alpha_{BP} = 1$ and $\beta_{BP} = 1$.)

The above-stated can be associated with the fact that light particles move relatively freely within the core sponge of enclosed carcass cells.

In polymers, the chain BP reaction, as well as the liquid-phase reaction, is inhibited by oxygen. Therefore, its order by BP is 1.0 (in contrast with the liquid-phase reaction order 1.5), and the material balance proves simple stoichiometry of BP + RH = rH + R-Ar, because the chain-carrying benzoyloxyls react without decarboxylation:

$$r^{\bullet} + RH \rightarrow rH + R^{\bullet},$$

$$R^{\bullet} + BP \rightarrow r^{\bullet} + R\text{-}Ar.$$

By this parameter CTA and PC films are distinguished from liquids.

In addition, simple stoichiometry of the chain reaction is coupled with that of the non-chain reaction: BP + Ph = rH + rPh + polymeric products.

Simultaneously, simple stoichiometry of the chain reaction is combined with the complex rate constant

$$k_{chain} = k_0 k_i C_0,$$

which includes the initial BP concentration. This first proves the existence of kinetic order one of the chain termination and, secondly, the competition of chain and non-chain ways of the common process proceeding. Besides, the efficiency of the chain reaction is practically independent of the chemical difference between CTA and PC. These facts cannot be explained by the homogeneous reaction mechanism which regulates the chain reaction order 1.5 as a result of quadratic law of the chain termination (Section 2).

The homogeneous reaction mechanism is also unable to explain why reactions of benzoyloxyls and macroradicals do not induce polymer chain breaks in the arylation reaction chains and why the breaks only are associated with the primary conversion of both benzoyloxyls and macroradicals.

So far as concerns the polymer degradation, the kinetic order one by BP (equations (3.3) - (3.5)) means that the primary macroradicals only undergo the polymer chain break and do not interact with either BP or $r^{\bullet}$. At the same time, secondary macroradicals produced by chain arylation make no breaks and only react, due to some reason, with BP molecules. The case is that two separated reaction chains must coexist and that the intermediates of the primary reaction chain do not affect those of the secondary one.

The noted peculiarities shall be explained taking into consideration specificity of the zonal structure of carcass-micellar microreactors which accumulate the BP additive.

3.3. Heterophase Mechanism of Polymer Reactions with BP

Taking into account that the force field of tight, or incapacious, nanopores prevents dissociation of both BP and $r^•$, it is assumed that these acts are practically inhibited in the intramicellar core sponge (*v*-zone). Simultaneously, dissociation is probable near *pc*-carcass domains in the capacious nanopores of the *s*-zones (supernanopores). Primary $PhCOO_s^•$ ($r_s^•$) radicals are transformed predominantly within the *s*-zones. This is associated with rather low molecular-segmental mobility of the polymer chains and light radicals reacting in the adsorbed state.

The layout of *s*-reactions, proceeding in supernanopores, presents stages of BP molecule interzone exchange as follows:

$$BP_v \xrightarrow{k_{vs}} BP_s,$$

$$BP_s \xrightarrow{k_{sv}} BP_v,$$

and the stage of the primary radical pair formation:

$$BP_s \xrightarrow{k_{0s}} \{2r_s^•\}. \qquad (3.0)$$

The radical pair either escapes from the cage:

$\{2r_s^•\} \rightarrow 2r_s^•$ (at the rate constant k_0')

or forms phenyl benzoate:

$\{2r_s^•\} \rightarrow rPh + CO_2$ (at the rate constant k_0'').

In CTA films, one share of escaped benzoyloxyl $r_s^•$ detaches those H atoms bound to CTA macromolecule backbones to form the free macroradical $P_s^•$ having a free valence in position 1 or 4 of glycopyranose rings (equation (3.I$_s$), $P_s^• \equiv$ >C$^•$–O–C<). The $P_s^•$ radicals are intermediates providing polymer chain breaks [32] (reaction (3.III$_s$)).

In PC films one share of escaped benzoyloxyl $r_s^•$ is captured by benzene rings to form the σ-complexes, playing the role of radicals $P_s^•$ which provide the ruptured polymer chains possessing tail cumene-type and phenoxyl-type groups [36, 37].

According to experiments, $P_s^•$ radicals do not take part in arylation processes, i.e. they do not react with BP and $r_s^•$. In these terms, the reactions can be written down as follows:

$r_s^• + PH_s \xrightarrow{k_{1s}'} rH + P_s^•$ (3.I$_{s,CTA}$)

(k_{1s}' $[PH]_s \equiv k_{1s}$ which characterizes CTA),

$$r_s^{\bullet} + PH_s \xrightarrow{k'_{1s}} P_s^{\bullet}\ (3.I_{s,CTA})$$

(k'_{1s} $[PH]_s \equiv k_{1s}$, σ-complex which characterizes PC).

Simultaneously, the rest part of escaped $r_s^{\bullet}$ radicals interacts with CTA and PC macromolecules to form the $R_s^{\bullet}$ free radicals which have free valences on the side groups of macromolecules:

$$r_s^{\bullet} + PH'_s \xrightarrow{k'_{2s}} rH + R_s^{\bullet} \qquad (3.II_s)$$

(k'_{2s} $[PH']_s \equiv k_{2s}$, both CTA and PC),

$$P_s^{\bullet} \xrightarrow{k_{3s}} R_{1s}^{\bullet} + \text{product} + \text{polymer chain break}, \qquad (3.III_s)$$

$R_s^{\bullet} \rightarrow$ termination products,

$R_{1s}^{\bullet} \rightarrow$ termination products

(the scheme of termination processes are below).

Applying the steady state condition to the present scheme of *s*-reactions, the expressions may be obtained:

$$-\frac{dC_s}{dt} = k_{0s}C_s;$$

$$[r_s^{\bullet}] = \frac{2\delta_i k_{0s} C_s}{k_{1s} + k_{2s}}; \qquad (3.10)$$

$$\left(\frac{d[rH]}{dt}\right)_{CTA} = (k_{1s,CTA} + k_{2s,CTA})[r_s^{\bullet}] = \delta_{CTA}k_{0s}C_s;$$

$$\left(\frac{d[rH]}{dt}\right)_{PC} = k_{2s,PC}\left[r_s^{\bullet}\right] = \frac{\delta_{PC}k_{0s}C_s k_{2s,PC}}{k_{1s,PC} + k_{2s,PC}};$$

$$\left(\frac{dn}{dt}\right)_{CTA} = \frac{\delta_{CTA}k_{0s}C_s k_{1s,CTA}}{k_{1s,CTA} + k_{2s,CTA}};$$

$$\left(\frac{dn}{dt}\right)_{PC} = \frac{\delta_{PC}k_{0s}C_s k_{1s,PC}}{k_{1s,PC} + k_{2s,PC}};$$

where $\delta_i = \dfrac{k'_{0,i}}{k'_{0,i} + k''_{0,i}}$.

With due regard to experimental results, macroradicals $R_s^{\bullet}$ and $R_{1s}^{\bullet}$ shown in the layout above must transform, forming no intermolecular chemical bonds. This is possible, if corresponded reactions proceed with participation of intermediate light free radicals. In CTA samples, the role of such radicals is played by acetyl radicals, produced from acetate groups:

$>CHOCOCH^{\bullet}_{2,s} \rightarrow >C^{\bullet}OCOCH_{3,s} \rightarrow >C{=}O + CH_3C^{\bullet}O_s$,

$CH_3C^{\bullet}O_s + R_s^{\bullet}, R_{1s}^{\bullet} \rightarrow$ stable products.

In PC samples these may be methyl radicals, synthesized in two ways:

1)

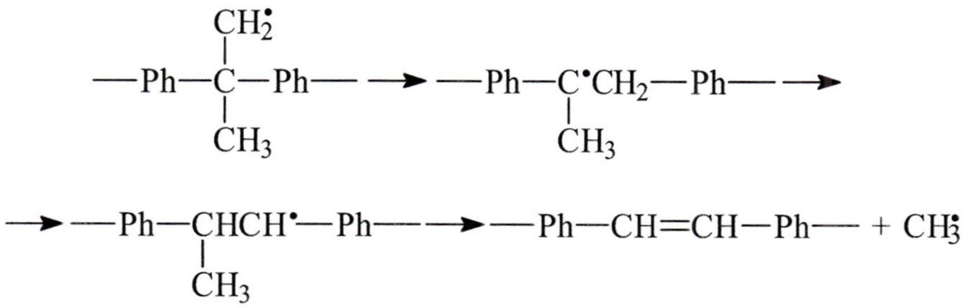

2)

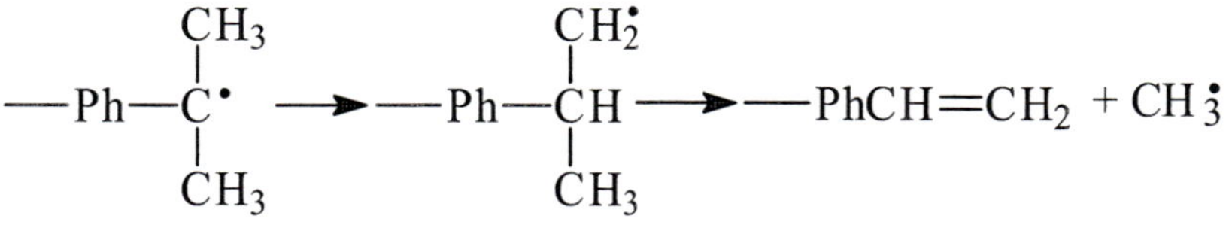

Note that considered *s*-reaction layout and obtained expression for concentration $[r_s^{\bullet}]$ do not assume free valence losses by means of their translation into *v*-zones of spongy micelles, in which reaction chains of BP decomposition are realized.

In line with the equation (3.2) and in accordance with the heterophase model (Section 3), kinetic description of the free radical *s*,*v*-translation necessitates supposition of its low rate which has no effect of the concentration $[r_s^{\bullet}]$. Moreover, it must be assumed that the *s*,*v*-translation itself is performed in the presence of mobile particles, promoting the interzone migration of free radicals. In this process, the promoter particles are low-molecular compounds with benzene rings (hereinafter, for the sake of simplicity denoted as ∅H). They are able to attach free radicals, forming σ-complexes which, in their turn, induce BP decomposition with free benzoyloxyls ($r^{\bullet}$) liberation (refer to Section 2). BP are classified as ∅H compounds, as well as benzoic acid (rH) and phenylbenzoate (rPh) obtained in the reaction.

The layout of the chemical reaction bridge, providing translation of free valences from *s*- to *v*-zones, is presented by reactions (3.IV) and (3.V) as follows:

$$r_s^{\bullet} + \varnothing H_s \rightarrow r_s\varnothing H^{\bullet}, \quad (3.IV)$$

$$r_s\varnothing H^{\bullet} + BP_s \rightarrow r_s\varnothing + rH + r^{\bullet}. \quad (3.V)$$

Benzoyloxyls $r^{\bullet}$, occurring in *v*-zones of spongy micelles due to the reaction (3.V), are impacted by torsional fields of incapacious nanopores. As a result, $r^{\bullet}$ stability is increased; dissociation

$$r^{\bullet} \rightarrow Ph^{\bullet} + CO_2$$

is eliminated, and probability of $r^{\bullet}$ reaction with CH-bonds of macromolecules increases. Within the incapacious *v*-nanopores, the conditions may be more favorable for $r^{\bullet}$ reactions with macromolecular side groups which may obtain more convenient reactive disposition during pulsation of the nanopores.

It is assumed that studied CTA and PC patterns possess nearly identical structural-physical characteristics of nanopores and intramicellar zones. The close similarity of these details is probably the reason for the equivalency of the rate constants of main stages in heterogeneous BP conversion. (In this connection, it should be noted that the globular elements of the spatial lattice have dimensions from 150 to 300 Å for both CTA [38] and PC [39].)

The above stated allows the authors to write a similar scheme of the chain reaction for CTA and PC:

$$r^{\bullet} + RH \xrightarrow{k_1} rH + R^{\bullet}, \quad (3.VI)$$

$$R^{\bullet} + BP_V \xrightarrow{k^i} RC_6H_4COOH + r^{\bullet}, RPh + CO_2 + r^{\bullet}, \quad (3.VII)$$

where $R^{\bullet}$ corresponds to the side radical $-^{\bullet}CH_2$. In both polymers, the arylation chain termination probably occurs according to the linear law, through $r^{\bullet}$ migration to *s*-zones followed by the reaction on the surface of the carcass cell:

$$r^{\bullet} \xrightarrow{k_2} r_s^{\bullet} \rightarrow rH + R_s^{\bullet}. \quad (3.VIII)$$

The recorded layout of the chain reaction does not correspond to the homogeneous version, as it levels the chemical difference between CTA and PC. Just this is required by the observed kinetic equivalence of both polymers in the chain reaction with BP.

Taking into account the exponential law of BP decomposition when initiating from a fixed C_0 value, the authors can assume the rate of the reaction not affecting the equilibrium of interzonal BP exchange. It is the condition that determines the steady concentrations C_s and C_v (i.e. $[BP_v]$):

$$\frac{dC_s}{dt} = k_v C_v - k_s C_s - k_{0s} C_s \approx k_v C_v - k_s C_s = 0,$$

whereof

$$C_s = \frac{k_v C_v}{k_s} = K_e C_v, \tag{3.11}$$

where K_e is the equilibrium constant.

Taking into account that v-nanopores capture the additive molecules more effectively than s-nanopores, the local concentrations C_s and C_v (equation (3.11)) may be reduced to the following relations [33]:

$$C_s \approx \frac{K_e C}{\alpha_{cel}}, \tag{3.12}$$

$$C_v \approx \frac{C}{\alpha_{cel}}, \tag{3.13}$$

where α_{cel} is the portion of spongy microreactors in the specimens.

It should be noted that direct proportionality between C_s, C_v and C will only be fulfilled in the absence of structural changes in reacting nanopores and under the stability of zonal masses in microreactors during the reaction.

Relations (3.12) and (3.13) allow expressing the theoretical law of BP decomposition in the polymers. Thus, the steady concentration of secondary benzoyloxyls $[r^{\bullet}]$ can be obtained when using the rate equality for initiation (3.IV) and termination (3.VIII) of arylation chains:

$$d[r^{\bullet}]dt = k_s[r_s^{\bullet}][\varnothing H_s] - k_2[r^{\bullet}] = 0,$$

taking into account the expression found for $[r_s^{\bullet}]$ (4.3.10):

$$[r^{\bullet}] = \frac{2\gamma\delta_i k_{0s} k_s C_0 C_s}{k_2(k_{1s} + k_{2s})},$$

where $[\varnothing H_s] = \gamma C_0$; γ is a constant which structure will be shown in Section 4.4.

Substituting $[r^{\bullet}]$ and $[r_s^{\bullet}]$ to the expression for the total rate of BP decomposition in spongy microreactors (W_{gs}), summed up from the rates of reactions (3.0) and (3.VII):

$$W_{gs} = -\left(\frac{dC_s}{dt} + \frac{dC_v}{dt}\right) = k_{0s} C_s + k^{i}[R^{\bullet}]C_v = k_{0s} C_s + k_1[r^{\bullet}][RH],$$

one obtains that

$$W_{gs} = k_{0s}C_s + [\mathrm{RH}]\frac{2\gamma\delta k_{0s}k_s k_1 C_0 C_s}{k_2(k_{1s} + k_{2s})}.$$

Thus taking into consideration the equations (3.12) and (3.13), the theoretical BP decomposition rate for the polymer films is obtained:

$$W = \alpha_{cel}W_{gs} = -\frac{dC}{dt} = k_0 + k_0 k_i C_0 C = kC, \tag{3.14}$$

where $k_0 = k_{0s}K_e$; k_i is the complex constant. The equation (3.14) is adequate to the empiric relations (3.1) and (3.2).

The other theoretical rates are as follows:

for rH formation:

$$W_{rH} = \alpha_{cel}\left(\frac{d[rH_s]}{dt} + \frac{d[rH_v]}{dt}\right) = \alpha_{cel}(k_{1s} + k_{2s})[r_s^{\bullet}] + \alpha_{cel}k_1[r^{\bullet}][RH] =$$

$$= (\alpha_{rH}k_0 + k_0 k_i C_0)C;$$

for arylated polymer (RAr) synthesis:

$$W_{RAr} = \alpha_{cel}\left(\frac{d[RAr_v]}{dt}\right) = \alpha_{cel}k_1[r^{\bullet}][RH] = k_i C_0 C;$$

for the polymer chain break:

$$W_n = \frac{dn}{dt} = \alpha_{cel}\frac{dn}{dt} = \alpha_{cel}k_{3s}[P_s^{\bullet}] = \alpha_{cel}k_{1s}[r_s^{\bullet}] =$$

$$= \frac{2\alpha_{cel}\delta k_{1s}k_{0s}C_s}{k_{1s} + k_{2s}} = \alpha_n k_0 C = k_n C, \tag{3.15}$$

also correspond to the experimental results, and the ratio $\beta_{rH} = \beta_{RAr} = 1$ is also valid.

It should be emphasized that in inferring the expressions above, the following inequalities were used:

$$k_s[r_s^{\bullet}][\varnothing H_s] << (k_{1s} + k_{2s})[r_s^{\bullet}];$$

$$k_s[r_s^{\bullet}][\varnothing H_s] << k_1[r^{\bullet}][RH],$$

which correspond to infrequent and long reaction changes in arylation.

3.4. Oxygen Effect in the Model BP Reaction

As has been noted, oxygen inhibits the chain stage in CTA and PC nearly to an equal extent (Figure 3, curve 2). In air, hydroperoxide polymers are not accumulated in the films, and a two-fold decrease in the degradation constant is observed:

$$k_n^a = 0.5k_n.$$

Pure oxygen does not change this situation.

This feature can be explained by the fact that oxygen only reacts with side and terminal free radicals (i.e. $R_s^{\bullet}$ and $R_{1s}^{\bullet}$ ones). In this reaction low-molecular $HO^{\bullet}$ radicals are produced. The latter interact with primary benzoyloxyls $r_s^{\bullet}$ that is the reason for two-fold decrease of the degradation constant k_n. For CTA and PC, radical $R_s^{\bullet}$ oxidation is described by the same scheme:

$$–CH_{2,s}^{\bullet} + O_{2,s} \rightarrow –CH_2OO_s^{\bullet} \rightarrow –C^{\bullet}HOOH_s \rightarrow –CH{=}O + HO^{\bullet}_s. \qquad (3.IX)$$

In these polymers, $R_{1s}^{\bullet}$ radicals are different, formation of free hydroxides also results their oxidation. For example, a polymer chain break in CTA (reaction (3.III$_s$)) gives terminal radical with a free valence at glucopyranose ring (of $>C^{\bullet}H$ type), and its oxidation proceeds similar to the reaction (3.IX):

$$>C^{\bullet}H_s + O_2 \rightarrow >CH(OO^{\bullet})_s \rightarrow >C^{\bullet}OOH_s \rightarrow >C{=}O + HO_s^{\bullet}.$$

Radical $P_s^{\bullet}$ decomposition in polycarbonate produces free phenoxyls and cumene radicals [36, 37]. Their oxidation reactions may be presented by the schemes, according to which $HO^{\bullet}$ radical liberation is combined with corresponding formation of end *ortho*-quinone and aldehydes groups:

$$^{\bullet}O–C_6H_{4,s}\sim + O_{2,s} \rightarrow O{=}C_6H_4(OO^{\bullet})_s\sim \rightarrow O{=}C_6H_3^{\bullet}(OOH)_s\sim \rightarrow$$

$$\rightarrow O{=}C_6H_3({=}O)_s\sim + HO_s^{\bullet};$$

$$–C_6H_4–C^{\bullet}(CH_3)_{2,s} + O_{2,s} \rightarrow –C_6H_4–C(CH_3)_2(OO^{\bullet})_{,s} \rightarrow$$

$$–C_6H_4–C(CH_3)(^{\bullet}CH_2)(OOH)_{,s} \rightarrow –C_6H_4–(CH_3)C–O \text{ (epoxide, E)} + HO_s^{\bullet},$$

\ /

CH_2

$$E \rightarrow –C_6H_4–(CH_3)CHCH{=}O.$$

Further on, radicals $HO_s^•$ formed in *s*-zones interact with benzoyloxyls $r_s^•$, giving peroxybenzoic acid, which molecules disproportionate with benzoic acid and oxygen release:

$$2C_6H_5COOOH_s \rightarrow 2C_6H_5COOH + O_2.$$

Such disproportionation reactions of peroxy acids and hydroperoxides proceed both in liquids and polymers [40, 41].

Based on the given description, one can easily deduce an expression for stationary benzoyloxyls' concentration in *s*-zones, containing dissolved oxygen, to replace the equation (3.10):

$$[r_s^•] = \frac{\delta_i k_{0s} C_s}{k_{1s} + k_{2s}}. \qquad (3.16)$$

This concentration is just twice lower versus that in the absence of oxygen (equation (3.10)). That is why in accordance with the experimental result, the rate constant of polymeric chain breaks (k_n^a) is twice lower, too (Section 3.1).

Thus it is important that the oxidation process must not involve synthesis of $P_s^•$ radicals, which should be decayed. Such an assumption seems to be reasonable with respect to steric conditions, owing to a decrease of O_2 interaction frequency with $P_s^•$ radicals, located on supernanopore walls. Side $R_s^•$ and terminal $R_{1s}^•$ radicals shall undergo oxidation more readily, as they appear in the volume of supernanopores during their vibration-rotation motions.

Inhibition of the BP chain decomposition can be explained by rapid oxidation of the radical σ-complexes [42, 43] in the *s*-zones:

$$r_s\varnothing H^• + O_{2,s} \rightarrow r_s\varnothing H(OO^•)_s \rightarrow r\varnothing + (HO_2^•)_s;$$

$$r_s^• + (HO_2^•)_s \rightarrow rH + O_{2,s},$$

when eliminating the interzonal *s,v*-transfer of free radicals.

At sufficiently high temperatures (378 - 388 K) a noticeable order one chain reaction of BP is observed in aerated CTA films [35]. Probably, in this case a part of low-active $HOO^•$ radicals attain the *v*-zones of microreactors, where they induce the chain reaction of BP decomposition:

$$HOO_s^• \rightarrow HOO_v^•,$$

$$HOO_v^• + BP_v \rightarrow rH + O_2 + r^•;$$

$$r^• + RH \rightarrow rH + R^•(-CH_2^•)_v,$$

$$-CH_2^•{}_v + O_2 \rightarrow (-C^•HOOH)_v \rightarrow (-CH{=}O + {}^•OH)_v \rightarrow -CH(OH)O_v^•,$$

$$-CH(OH)O_v^• + BP_v \rightarrow -C(O)OH + rH + r^•,$$

$r^{\bullet} + RH \rightarrow rH + R^{\bullet}$.

Here alcohol ν-radical ($–CH(OH)O_v^{\bullet}$) is the carrier in the kinetical chain propagation.

Notation. Structural-kinetic model of the BP reaction was designed with the consideration of supramolecular NCPM structures, which are absent in the homogeneous system and characterized by the nanoheterogeneous scale. As is known for heterogeneous systems, the rate of chemical reactions is in many respects dependent upon the structural rearrangement factor. The present layout is related to polymeric specimens with stable carcass-micellar structure. Meanwhile, the structure of micro- and nanoheterogeneous elements in the system is capable of changing under thermomechanical effects, then changing kinetics of the complex reaction.

Actually, the experiments show significant changes in the characteristic function $k = f(C_0)$ of BP decomposition for CTA films, prepared using water-saturated methylene chloride as a solvent, and for CTA fibrous wool obtained by polymer precipitation from the solution [6]. The BP reaction kinetics change considerably when the PC films are drawn for a structural evolution [7], namely, the decomposition rate increases considerably, probably due to enhanced mobility of the elements of structural zones and increased frequency of the interzonal transfer of free radicals.

Thus under the supramolecular structure conditions, different polymers (such as CTA and PC) must simulate each other in a reaction and, vice versa, a polymer possessing an original molecular structure must simulate different chemical compounds. Shortly, kinetics of the polymer chain reaction is affected by both prehistory of the specimens and their structural change in the course of the reaction.

The next Section is devoted to photochemical BP reaction with CTA in the temperature range from 293 to 341 K. In this reaction, the structure of the films, obtained in the case of dry methylene chloride, should be quite stable.

3.5. Kinetic Model of BP Photodecomposition

The photochemical BP interaction with CTA (in the temperature range of 293 - 341 K) has been studied when using UV radiation with the wave numbers $\upsilon < 32{,}000$ cm^{-1} (which falls outside the region of the self CTA absorption) [44]. At C_0 values under consideration, optical density ($D < 0.05$) of the films provided the UV absorption nearly uniformly upon the specimen thickness (15 - 25 μm). The kinetics of the photochemical process is described by an exponential equation similar to the thermal one (3.1). In evacuated films, the effective rate constant of BP photodecomposition

$$k_p = k_{0,p} + k_{i,p}C_0 \tag{3.17}$$

is a higher value (Figure 8, curve 1), than in the films aerated by ambient air (Figure 8, curve 2):

$$k_p^a = k_{0,p}^a + k_{i,p}^a C_0,$$

where $k^{a}_{0,p}$ = $k_{0,p}$ is the photodissociation rate constant, the quantum yield of photodissociation < 0.1, and the activation energy $E_{0,p} \approx 13$ kJ/mol; $k^{a}_{i,p} \approx 0.5k_{i,p}$.

Kinetic phenomenology of the photoprocess does not differ qualitatively from the thermal one. Accordingly, it may be assumed that the above-discussed heterophase mechanism is suitable for the BP photo decomposition as well. As a consequence, BP molecules, absorbing UV radiation, are thought to be only dissociated in the *s*-zones of spongy micelles. At the same time, the BP photodissociation should be inhibited by a force field in incapacious *v*-nanopores, thus in the *v*-zones, the energy of BP photoexcitation should be converted into heat only.

The above-mentioned means that the photoprocess must include the interzonal *s,v*-transfer of free valence. The *s,v*-phototransfer only may be associated with a specific phenomenon [44] when exposing BP doped films to UV-band with υ_1 = 30,500 cm^{-1}: the BP decomposition rate increases under simultaneous exposure to the second light source irradiating UV component with low wave numbers υ_2 < 27,600 cm^{-1}, insufficient for BP decomposition. In experiments [44], the k_p value changed from 6.5×10^{-4} to 8.5×10^{-4} s^{-1}, and the k^{a}_{p} value changed from 3.7×10^{-4} to 4.7×10^{-4} s^{-1} at such (two) light sources.

This effect evidences for intermediate photosensitive particles, which steady concentration is achieved from the beginning of the conversion. By spectral region of photosensitivity, these particles correspond to benzene σ-complexes (undergoing photolysis with H atom ejection [45]) and oxidation to corresponded peroxyls [42] (refer to Section 1).

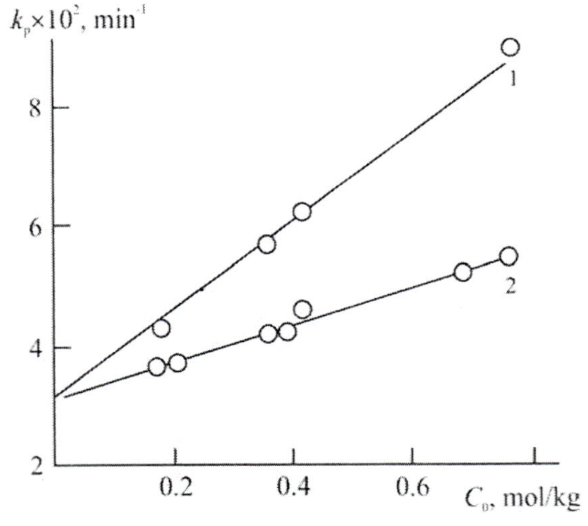

Figure 8. Dependence of effective constant of the BP photolysis rate on C_0 in evacuated (1) and aerated (by ambient air) (2) CTA films at 293 K. UV-radiation intensity in the spectrum region of υ< 32,200 cm^{-1} is 7.6×10^{-3} einstein·m^{-2}s^{-1}

Taking into account the above consideration, one may assume thermal *s,v*-transfer of the radicals (reactions (3.IV) and (4.3.V)) becoming practically 'frozen' at T < 340 K (it occurs at low frequency at 371 K). Meanwhile, the possibility of phototransfer remains due to photolysis of the σ-complexes and their peroxyls. The *s,v*-transfer scheme for the evaluated films will attain the following form:

$$r_s^{\bullet} + \varnothing H_s \xrightarrow{k_s} r_s\varnothing H^{\bullet} \xrightarrow{k_{2,p}(\upsilon_1 \div \upsilon_2)} r\varnothing + H_v^{\bullet},$$

$$r_s\varnothing H^{\bullet} \xrightarrow{k_{-s}} r_s^{\bullet} + \varnothing H_s,$$

$$H_v^{\bullet} + BP_v \rightarrow rH + r^{\bullet}.$$

In retaining other elementary *s*- and *v*-stages of the heterophase mechanism (Section 3.3), this scheme allows obtaining of the kinetic equation of BP photodecomposition.

Taking into account equality of the initiation and the termination rates for the reaction chains in the *v*-zone:

$$k_{2,p}[r_s\varnothing H^{\bullet}] = k_2[r^{\bullet}]$$

and expressing the term $[r_s\varnothing H^{\bullet}]$ in compliance with the *s,v*-phototransformation scheme, it is found that

$$[r^{\bullet}] = \frac{k_{2,p}k_s[r_s^{\bullet}][r_s\varnothing H^{\bullet}]}{k_2(k_{-s} + k_{2,p})}. \tag{3.18}$$

Then substituting the expression obtained for $[r^{\bullet}]$ into the equation for BP photo decomposition rate

$$W_{p,gs} = k_{0s,p}C_s + k_1[r^{\bullet}][RH]$$

and taking into consideration the equations (3.10), (3.12) and (3.13), the following theoretical correlation is deduced:

$$\frac{dC}{dt} = \left[k_{0,p} + const\frac{k_{2,p}k_{0,p}C_0}{k_{-s} + k_{2,p}}\right]C = (k_{0,p} + k_{i,p}C_0)C = k_pC, \tag{3.19}$$

where constant k_p is identical to the experimental one (3.16).

According to the theoretical equation (3.19), $k_{i,p}$ value displays superlinear dependence on the intensity (*I*)of acting UV radiation, as the numerator in $k_{i,p}$ includes the product of $k_{0,p}$ and $k_{2,p}$ constants which depend on *I*. This should be observed for low $k_{2,p}$ versus k_{-s} in the reverse dissociation of $r_s\varnothing H_s^{\bullet}$ radicals. Actually, a superlinear dependence is observed [44].

The equation (3.19) structure indicates the reason for intensification of BP decomposition with a supplementary source of long-wave radiation with $\upsilon_2 < 27{,}600\ cm^{-1}$. This radiation increases the rate constant $k_{2,p}$ of $r_s\varnothing H_s^{\bullet}$ radical decomposition toward the chain stage initiation of in *v*-zones of the carcass spongy micelles.

The efficiency of the chain decomposition $\left(\frac{k_{i,p}}{k_{0,p}}\right)$ calculated by the data, shown in Figure 4.8, is close to $\frac{k_c}{k_0}$ value, calculated from Figure 3 (at T = 371 K). Probably, at room temperature the *s*,*v*-phototransfer of free radicals is combined with a high translational mobility of BP and $r^{\bullet}$ particles within spongy micelle cores of the polymer supramolecular carcass.

So far as concerns the films aerated in ambient air, one can assume oxidative *s*- and *v*-reactions proceeding independently of photoreaction of $r_s\varnothing H_s^{\bullet}$ radicals, remaining unoxidized. Under this condition, $r_s^{\bullet}$ concentration is a half of that in evacuated films:

$$[r_s^{\bullet}] = \frac{\delta_i k_{0s} C_s}{k_{1s} + k_{2s}}. \tag{3.16}$$

Assuming that the reaction chains in the *v*-zone are presented by the scheme:

$H_v^{\bullet} + O_2 \rightarrow HOO_v^{\bullet}$;

$HOO_v^{\bullet} + BP_v \rightarrow r^{\bullet} + rH + O_2$;

$r^{\bullet} + -CH_3 \rightarrow rH + -CH_2^{\bullet}$;

$-CH_2^{\bullet} + O_2 \rightarrow -CH_2OO^{\bullet} \rightarrow C^{\bullet}HOOH \rightarrow -CH(OH)O^{\bullet}$;

$-CH(OH)O^{\bullet} + BP \rightarrow -COOH + rH + r^{\bullet}$,

and their termination happening by the former reaction (3.VIII), it is found that

$$[r^{\bullet}]^a = 0.5[r^{\bullet}].$$

Taking into consideration expressions (3.12), (3.13), (3.16) and (3.18), the following expression of the BP photodecomposition rate for the air-aerated films is obtained:

$$\frac{dC}{dt} = \left[k_{0,p} + const^a \frac{k_{2,p} k_{0,p} C_0}{k_{-s} + k_{2,p}}\right] C = \left(k_{0,p} + k_{i,p}^a C_0\right) C = k_p^a C, \tag{3.20}$$

which agrees with the experiment, where $k_{i,p}^a$ = $0.5k_{i,p}$. According to the equation (3.20), long-wave UV radiation (non-absorbable by BP at $\upsilon_2 < \upsilon_1$) can increase the reaction rate under actually observed condition [42].

Completing this Section, note that the mechanism of BP photoconversion was not previously formulated [44], because the discussion logics did not go beyond the frames of the homogeneous scheme. As shown in the Section, the successful result is achieved due to taking into consideration the functions of nanoheterogeneous elements of the supramolecular structure during the chemical process.

4. Heterophase Mechanism of Polystyrene Arylation by Dibenzoyl Peroxide

4.1. Phenomenological Features of the Process

Among the first investigators of the reaction in question were J.W. Breitenbach and H. Frittum, who have implemented experiments with chlorine-substituted BP derivatives [46]. Studying evacuated PS films, they have come up against an abnormal behavior of free benzoyloxyl radicals. These radicals are known to readily decompose in liquids to free phenyl radicals and carbon dioxide. However, in some experiments [46] they were efficiently linked to polymer chains in amount corresponding to 40 - 50% of added BP fragments. The remaining BP fragments attached themselves to macromolecules, shaped as chlorophenyl groups.

The experiments with aerated PS films implemented by H.C. Haas [47-49] revealed a great stability of benzoyloxyl radicals. For the products of conversion, Haas has identified benzoic acid, phenylbenzoate, and polymeric benzoates. The latter displayed higher output than in [46] and BP phenyl groups attached to polymer chains were not detected. In accordance with the Haas data, benzoyloxyl radicals do not dissociate in the polymeric medium and are bound to benzene rings and secondary but not ternary carbon atoms of the polymer chain.

Macroradicals originated in aerated PS films underwent decomposition, accompanied by cleavage of the polymer chains and formation of low-molecular fragments with unsaturated C=C bonds. It seemed strange that alkyl macroradicals did not react with oxygen dissolved in the films. Despite free air diffusion into the films, the chemical analysis indicated complete absence of the oxidation products. This property distinguishes PS from its liquid chemical analog – isopropylbenzene, which is oxidized under BP initiation, thus producing a hydroperoxide with high output [50].

A comprehensive kinetic investigation undertaken in [47 - 49] has elucidated a non-chain origin of the BP reaction in aerated PS films. At the same time, the reaction rate proved to exhibit an essentially non-Arrhenius temperature dependence. Actually, at 353 K the activation energy E0 = 126 kJ/mol, meantime at a lower temperature of 329 K that increases to 151 kJ/mol. Haas ascribed the abnormal tendency to a strong cage effect, however, this explanation ran counter to another fact revealed by his experiments that the quantitative composition of the reaction products was temperature-independent.

R. Rado and M. Lazar [2 - 5], experimenting with the same systems, have found that rmoval of oxygen from the films entails a chain reaction of BP. However, they saw no need to analyze the kinetics of this reaction more closely. This was later done by V.P. Pustoshnyi [51, 52].

According to [52] (experiments with a polymer of molecular mass Mn = 150,000 and 15 - 20 μm thick films [51]), the kinetics of BP decomposition in the films under conditions of thermal and photochemical generation of radicals obeys the exponential law

$$-\frac{dC}{dt} = kC;\ C = C_0\exp(-kt), \tag{4.1}$$

where C0 and C are the initial and current BP concentrations; k is the effective rate constant dependent on C0. For the thermal reaction, the dependence of the rate constant k on concentration C0 over the temperature range from 353 to 372 K is expressed by an equation

$$k = k_0 + \frac{k_0 a C_0}{1 + bC_0}, \tag{4.2}$$

which is similar in appearance to that for polyethylene and poly(ethylene oxide) [53, 54] (see Sections 1, 3). The first and second terms in (4.2) are the rate constants for the non-chain and chain reactions, respectively, and *a* and *b* are coefficients.

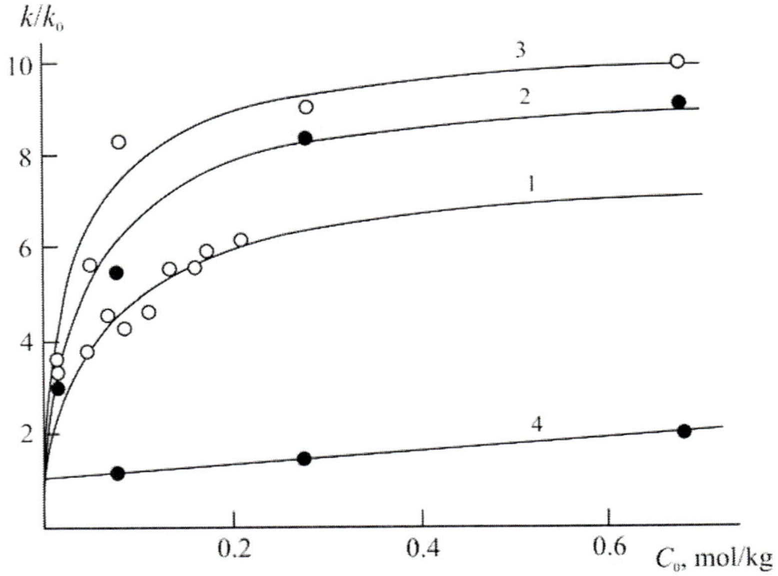

Figure 9. Effective rate constant of BP decomposition in evacuated PS films as a function of initial BP concentration C_0 at temperature (1) 353, (2) 363, (3) 372, and (4) 299 K; line (4) corresponds to the photochemical process induced by filtered UV-radiation in narrow spectral region with λ_{max} = 313 nm

Plots 1 - 3 (Figure 9) of the dependence

$$\frac{k}{k_0} = 1 + \frac{aC_0}{1 + bC_0}, \tag{4.3}$$

stemming from equation (4.2) are seen to agree reasonably well with the experiment. The curves in Figure 9 were calculated with $k_0 \times 10^5$ = 1.11; 3.33; 10.0 s^{-1} (these constants were borrowed from [45 - 47]); a = 95; 155; 280 kg/mol at 353, 363, and 372 K, respectively.

The points for curve 1 (Figure 9, 353 K) were borrowed from [4] where they were derived from the degrees of BP decomposition at given times and C_0, other points been borrowed from [54]. Note that the above-indicated coefficients a and b correspond to approximately equal activation energies $E_a \approx E_b \approx 59$ kJ/mol.

As is known, the chain BP reaction in liquid aromatic hydrocarbons proceeds slower than in PS [51, 52] possessing the order 1.5 (Section 2). Experiments on BP decomposition in deaerated toluene in sealed glass ampoules, accomplished by Pustoshnyi for comparison, have led to the conclusion that the reaction rate at 372 K obeys a commonplace liquid-phase equation

$$-\frac{dC}{dt} = k_{o1} + k_1 C^{1.5},$$

where $k_{o1} = (3.0 \pm 0.2)\times10^{-4}\ \text{s}^{-1}$ and $k_1 = (2.4 \pm 0.5)\times10^{-4}\ (\text{l/mol})^{0.5}\text{s}^{-1}$.

It turns out that the original efficiency of chain decomposition at $C_0 = 0.25$ mol/kg in liquid toluene is low, $\frac{k_1 C^{0.5}}{k_{01}} = 0.4$, compared to $\frac{k - k_0}{k_0} = 8.0$ in a polymer which possesses an exceptionally high macroscopic viscosity. This implies BP molecules in the polymer residing in structural zones where they move almost as freely as in liquids.

BP molecules also exhibit high mobility at lower temperatures of 299 and 313 K. This is evidenced by a tangible contribution of the chain component to the rate of photochemical decomposition (line 4 in Figure 1). In this case, $k(C_0)$ represents a straight line similar in other glassy polymers: polycarbonate, cellulose triacetate (Section 3), and polyamide PA-548 (Section 2).

4.2. Polystyrene Arylation Mechanism

As follows from the polymer microreactor model, initiated conversion of a polymer proceeds via exchange of dibenzoyl peroxide molecules between v- and s-zones of spongy micelles

$$\text{BP}_\text{v} \xrightarrow{k_{vs}} \text{BP}_\text{s};$$

$$\text{BP}_\text{s} \xrightarrow{k_{sv}} \text{BP}_\text{v},$$

where subscripts ‘v’ and ‘s’ imply that the molecule resides in v- and s-zones, respectively. The exchange process rapidly comes to equilibrium

$$k_{vs}C_v = k_{sv}C_s \text{ or } C_s = \frac{k_{vs}}{k_{sv}}C_v = \frac{K''}{\alpha_{cel}}C, \qquad (4.4)$$

where C_s, C_v, and C are BP concentrations in *s*- and *v*-zones, and in entire sample; α_{cel} is the weight fraction of the globular spongy micelles; $K'' = \frac{k_{vs}}{k_{sv}} < 1$ is the constant of equilibrium between the zones. The equilibrium cannot be disturbed by the reaction for the reaction proceeds much slower than BP molecules migrate in the nanopores of spongy micelles that are the real polymer microreactors.

As noted above, dissociation acts only occur in the *s*-zones and give rise to primary radical pairs

$$BP_s \xrightarrow{k_{0s}} \{2r_s^{\bullet}\},$$

where $r^{\bullet} = PhCOO^{\bullet}$.

Some of these pairs transform to free radicals:

$$\{2r_s^{\bullet}\} \xrightarrow{k'_{0s}} 2r_s^{\bullet},$$

and some combine into phenylbenzoate

$$\{2r_s^{\bullet}\} \xrightarrow{k''_{0s}} CO_2 + PhCOOPh\ (Phr).$$

As benzoyloxyl radicals become free, they react with PS to detach hydrogen atoms from tertiary carbon atoms of polymer chains

$$r_s^{\bullet} + \text{–(Ph)CH}_{tert,s}\text{–} \xrightarrow{k'_{1s}(CH_{t,s})} rH + \text{–C}^{\bullet}\text{(Ph)CH}_2\text{–}\ (P_s^{\bullet};\ (k'_{1s}\,[CH_{t,s}] \equiv k_{1s}),$$

and the nascent backbone radical $P_s^{\bullet}$ decomposes with a polymer chain break to fragments with unsaturated terminal bonds and groups ($R_s^{\bullet}$) possessing free valences

$$P_s^{\bullet} \xrightarrow{k_{2s}} \text{–C(Ph)} = CH_2 + PhC^{\bullet}H\text{–}\ (R_s^{\bullet}),$$

which, in turn, combine with benzoyloxyl radicals

$$R_s^{\bullet} + r_s^{\bullet} \xrightarrow{k_{3s}} PhCH(OCOPh)\text{–}\ (Rr).$$

Furthermore, radicals $r_s^{\bullet}$ add themselves to phenyl rings of PS

$$r_s^{\bullet} \xrightarrow{k'_{4s}(Ph_s-)} rC_6{}^{\bullet}H_{5,s}\text{–}\ (\,k'_{4s}\,(Ph_s\text{–}) \equiv k_{4s}),$$

$$r_s^{\bullet} + rC_6{}^{\bullet}H_{5,s}\text{–} \rightarrow rH + rC_6H_4\text{–}\ (Pr).$$

The *s*-nanopore volume is sufficiently large to support the above reaction (a transformation from the planar to bent conformation of the phenyl ring attached to *s*-segment) may occur.

This *s*-mechanism explains the routes of formation of all the aerobic-arylation products identified by Haas, namely, of benzoic acid, phenylbenzoate (Phr), broken macromolecules, and two kinds of PS benzoates, Pr and Rr. Decarboxylation of benzoyloxyl radicals $r_s^{\bullet}$ was omitted from the mechanism, because these radicals spend the largest share of their lifetime in the adsorbed state on segments of *s*-nanopores (like any species in supernanopores).

Oxygen attachment to free *s*-radicals of PS is also missing in the reaction *s*-scheme. Experiments [47 - 49] revealed no oxidation of alkyl macroradicals, which was attributed to the structural factor. The point is that O_2, as any other low-molecular compound, is predominantly accumulated in narrow nanopores of *v*-zones. Probably, the inevitable hindrances in a supernanopore accompany steric conditions which are unfavorable for interaction of $P_s^{\bullet}$ and $R_s^{\bullet}$ radicals formed on rigid *s*-segments with an O_2 molecule which remains adsorbed on the pore walls for much of its residence time in the pore. Consequently, the probability of $P_s^{\bullet}$ and $R_s^{\bullet}$ conversions concurrent to oxidation increases dramatically.

The dynamics of quasi-solid *s*-segments of a spongy carcass is such that almost all *s*-radicals are converted without leaving their zones, and free-valence transfer to *v*-zones of the sponge is rather hindered and occurs only with a help of mobile particles.

In the case at issue, the mobile species promoting valence transfer are BP and also phenyl benzoate and benzoic acid arising by BP decomposition. All these species (∅H) possess phenyl nuclei and can transform into *σ*-complexes through addition of benzoyloxyl radicals

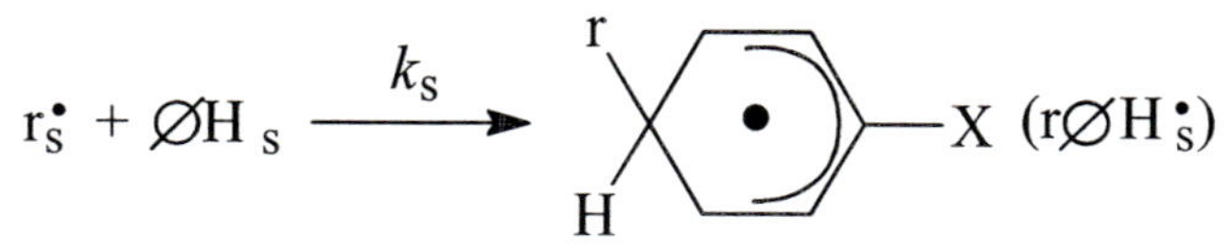

σ-Complexes, in turn, can induce BP decomposition

$$r\varnothing H_s^{\bullet} + BP_s \rightarrow r\varnothing + rH + r^{\bullet},$$

and under a suitable disposition of the attacking BP molecule in an *s*-nanopore, a free valence is transferred to a *v*-zone by the secondary (mobile) benzoyloxyl radical $r^{\bullet}$, despite a low probability.

When migrating in micropores of *v*-zones, benzoyloxyl radicals detach hydrogen atoms from tertiary carbon atoms of the polymer chains

$$r^{\bullet} \xrightarrow{k_1'(CH_{tert})} rH + P\bullet \ (k_1'(CH_{tert}) \equiv k_1)$$

or undergo decarboxylation

$$r^{\bullet} \xrightarrow{k_2} Ph^{\bullet} + CO_2,$$

$$Ph^{\bullet} \xrightarrow{k_3'(CH_{tert})} C_6H_6 + P^{\bullet} \; (k_3'(CH_{tert}) \equiv k_3).$$

The decarboxylation reactions of $r^{\bullet}$ radicals require much less activation energy than peroxide bond cleavage in BP, and volume fluctuations of *v*-nanopores are capable to facilitate these. Note the fluctuations of nanopores arranged from flexible PS-chains are more frequent as compared with those of PC and CTA distinguishing more rigid polymer chains. Changing in size, fluctuation-induced nanopores put these reactions in unequal conditions. For example, in fluctuationally low-active pores, the probability of $r^{\bullet}$ interaction with the nanopore walls to yield benzoic acid (rH) must be higher than in fluctuationally active nanopores where the prevailing $r^{\bullet}$ reaction is decarboxylation. In support of this hypothesis, chain decomposition of BP in glassy polycarbonate and cellulose triacetate primarily yields benzoic acid, presumably because rigid links of these polymers limit fluctuations of *v*-nanopores in amplitude and frequency of certain sizes.

Features of the spongy skeleton of *v*-zones and the smallness of their *v*-nanopores let the authors omit the items:

1. $r^{\bullet}$ and $Ph^{\bullet}$ addition to phenyl rings of PS (these reactions are known to bend the rings);
2. rupture of polymer chains by transformation of $P^{\bullet}$ radicals. Radicals $P^{\bullet}$ most frequently interact with migrating BP molecules, thus inducing chain arylation of PS:

$$P^{\bullet} + BP_v \xrightarrow{k_i} PC_6H_4COOH + r^{\bullet}, PPh + CO_2 + r^{\bullet}$$

(this reaction is presented by analogy with liquid hydrocarbon arylation).

Reaction *v*-chains are terminated in two ways, first, by migrating of mobile radicals to *s*-zones

$$r^{\bullet} \xrightarrow{k_4} r_s^{\bullet},$$

and, secondly, by attacking phenyl rings of ∅H compounds in *v*-nanopores

$$r^{\bullet} + \varnothing H_v \xrightarrow{k_5} r\varnothing H^{\bullet};$$

$$r^{\bullet} + r\varnothing H^{\bullet} \rightarrow rH + r\varnothing.$$

This complete conversion model, having all the basic features of the heterogeneous mechanism, yields the reaction rate in a unit cell of a supramolecular skeleton in the form of a sum of local BP decomposition rates in the individual nanophases. Its components are specific rates of the *s*-reaction

$$-\frac{dC_s}{dt} = k_{0s}C_s \tag{4.5}$$

and *v*-reaction

$$-\frac{dC_v}{dt} = k_i[P^\bullet]C_v = (k_1 + k_2)[r^\bullet] \quad (4.6)$$

derived from the foregoing *s*- and *v*-schemes.

The steady-state concentration of *v*-benzoyloxyl radicals is calculated from the equality of initiation and termination rates of reaction chains:

$$[r^\bullet] = \frac{k_s\left[r_s^\bullet\right]\left[\varnothing H_s\right]}{k_4 + 2k_5\left[\varnothing H_s\right]},$$

and that of *s*-benzoyloxyls results from the *s*-reactions scheme

$$[r^\bullet] = \frac{\delta k_{0s} C_s}{k_{1s} + k_{4s}}, \quad (4.7)$$

where $\delta = \dfrac{k'_{0s}}{k'_{0s} + k''_{0s}}$.

Substitution of $r^\bullet$ and $r_s^\bullet$ concentrations into equation (4.6) results in the specific rate of BP chain decomposition in the *v*-zones

$$-\frac{dC_v}{dt} = \frac{(k_1 + k_2)k_s \delta k_{0s}\left[\varnothing H_s\right]C_s}{(k_{1s} + k_{4s})(k_4 + 2k_5\left[\varnothing H_s\right])}. \quad (4.8)$$

There is a relation between the experimental and local rates, which can be deduced from the definitions of the concentrations:

$$C_s = \frac{n_s}{m_s};$$

$$C_v = \frac{n_v}{m_v};$$

$$C = \frac{n}{m} = \frac{n_s}{m} + \frac{n_v}{m} = \frac{m_s C_s}{m} + \frac{m_v C_v}{m} \approx \frac{m_v C_v}{m} \approx \alpha_{cel} C;$$

here, n_s, n_v, and n are the numbers of BP moles in the *s*- and *v*-zones and in the entire sample; m_s, m_v, and m are the masses of the *s*- and *v*-zones and the sample;

$$\alpha_{cel} = \frac{m_s + m_v}{m} \approx \frac{m_v}{m}.$$

Hence, the net rate of BP decomposition equals

$$-\frac{dC}{dt}=-\frac{m_s}{m}\frac{dC_s}{dt}-\frac{m_v}{m}\frac{dC_v}{dt}=-\alpha_{cel}\frac{m_s}{m_v}\frac{dC_s}{dt}-\alpha_{cel}\frac{dC_v}{dt},$$

which, as recast in terms of equations (4.5) and (4.8), becomes as follows:

$$-\frac{dC}{dt}=\alpha_{cel}\frac{m_s}{m_v}k_{0s}C_s+\frac{\alpha_{cel}(k_1+k_2)k_s\delta k_{0s}[\varnothing H_s]C_s}{(k_4+2k_5)[\varnothing H_s](k_{1s}+k_{4s})}. \quad (4.9)$$

When expressing the local concentrations C_s, $[\varnothing H_s]$, and $[\varnothing H_v]$ appearing in the equation (4.9) in terms of the dibenzoyl peroxide concentration C, it should be taken into account that some phenyl nuclei of BP are grafted in the polymer in the course of the reaction and the equivalent number of mononuclear particles are liberated. Thus the initial concentration of mobile ∅H molecules remains invariant, and almost all of them are localized in *v*-zones:

$$C_v+[Phr_v]+[rH_v]+[C_6H_{6,v}]=C_{0v}\approx\frac{C_0}{\alpha_{cel}}. \quad (4.10)$$

Local concentrations of phenyl nuclei of the low molecular ∅H compounds are related to the molecular concentration as follows:

$$[\varnothing H_s]=2C_s+2[Phr_s]+[rH_s]+[C_6H_{6,s}], \quad (4.11)$$

$$[\varnothing H_v]=2C_v+2[Phr_v]+[rH_v]+[C_6H_{6,v}]. \quad (4.12)$$

It is reasonably assumed that binuclear particles migrate in *v*-nanopores twice as slowly as mononuclear ones. This statement can be formulated in terms of the ratios of the rate constants and equilibrium distributions of the (4.4) type as follows:

$$k'_{vs}=2k''_{vs};\ k'_5=2k''_5;\ K'=2K'';$$

$$\frac{C_s}{C_v}=\frac{[Phr_s]}{[Phr_v]}=K''; \quad (4.13)$$

$$\frac{[rH_s]}{[rH_v]}=\frac{[C_6H_{6,s}]}{[C_6H_{6,v}]}=2K''.$$

Here the number of upper indices corresponds to the number of phenyl nuclei in a molecule. In terms of expressions (4.13), the equation (4.11) comes to the shape as follows:

$$[\varnothing H_s] = 2K''[C_v + [Phr_v] + [rH_v] + [C_6H_{6,v}]), \tag{4.14}$$

which by virtue of equation (4.10) yields

$$[\varnothing H_s] = 2K''\frac{C_0}{\alpha_{cel}}. \tag{4.15}$$

Equations (4.14) and (4.15) imply that the activity of phenyl nuclei in transferring a free valence between the zones remains invariant, despite a decrease in the total content of phenyl nuclei due to their grafting on PS.

In addition, the change of binuclear BP molecules by more mobile mononuclear molecules in the course of the reaction maintains the reactivity of ∅H-containing compounds in *v*-chain termination reactions at a constant level. Actually, the term $k_5[\varnothing H_v]$ in the denominator of equation (4.9) can be recast in terms of equations (4.12), (4.13), and (4.10) as

$$k_5[\varnothing H_v] = 2k_5''(C_v + [Phr_v] + [rH_v] + [C_6H_{6,v}]) = \frac{2k_5''C_0}{\alpha_{cel}}. \tag{4.16}$$

Substituting equations (4.4), (4.15), and (4.16) into equation (4.9), eventually the expession for the overall reaction rate in the whole sample is obtained:

$$-\frac{dC}{dt} = k_0C + \frac{k_0aC_0C}{1+bC_0}, \tag{4.17}$$

where

$$k_0 = k_{0s}\frac{K''m_s}{m_v},\ a = \frac{2(k_1+k_2)k_s\delta K''m_v}{(k_{1s}+k_{4s})k_4\alpha m_s},\ b = \frac{4k_5''C_0}{\alpha_{cel}k_4}.$$

Equation (4.17) guides the physical meaning of empirical equations (4.1) - (4.3) and their coefficients. The non-Arrhenius behavior of the effective rate constant k_0, noted by Haas [47 - 49] (as well as *a* and *b*) presumably owes itself to the structural factor affecting K'' (and k_5''), namely, to the temperature dependence of the nanopore size in a polymeric sponge.

The proposed mechanism reconciles the experimental data of [46] with the results of [47 - 49]. The point is that the process of BP decomposition in [46] involved both *s*- and *v*-reactions, because the test samples contained no dissolved oxygen. In contrast to [46], the samples in [47 - 49] were accessible to oxygen inhibiting the chain reaction [4, 5].

As is known, oxygen also inhibits BP chain decomposition in liquid solutions [8, 31, 50], because it oxidizes alkyl radicals, and the nascent free peroxyls are unreactive to BP. However, since the macroradicals observed in Haas's experiments underwent no oxidation, it can be concluded that the inhibition mechanism in PS matrices is other than in the liquid phase. In this case, free valence transfer between the *s*- and *v*-zones is presumably off, because σ-complexes mediating this transfer undergo an oxidation process which induces free radical termination

$$r\varnothing H_s^{\bullet} + O_{2,s} \rightarrow r\varnothing + HO_{2,s}^{\bullet};$$

$$r_s^{\bullet} + HO_{2,s}^{\bullet} \rightarrow rH + O_2.$$

In this case, chain processes converting BP to benzene and benzoyl acid are suppressed and the yield of PS benzoates builds up.

The local BP decomposition rate in the Haas experiments must obey the equation (4.5). The reaction rate in the entire sample equals

$$-\frac{dC}{dt} = -\alpha_{cel}\frac{m_s}{m_v}\frac{dC_s}{dt} = \frac{m_s}{m_v}k_{0s}K''C = k_0 C,$$

which yields upon integration of the exponential kinetic law

$$C = C_0\exp(-k_0 t).$$

According to the *s*-mechanism valid under the aerobic conditions, the yields of such reaction products as PS benzoate, benzoic acid, and phenylbenzoate per dibenzoyl peroxide decomposed in time *t* ($\Delta C_t = C_0[1 - \exp(-k_0 t)]$) are:

$$\frac{([\mathrm{Pr}]+[\mathrm{Pr}])_t}{\Delta C_t} = \frac{m\delta}{m_s},$$

$$\frac{[rH]_t}{\Delta C_t} = \frac{m\delta k_{1s}}{(k_{1s}+k_{4s})m_s},$$

$$\frac{[Phr]_t}{\Delta C_t} = \frac{m(1-\delta)}{m_s}.$$

There are no the rate constant k_{0s} of BP dissociation and the equilibrium constant K'' of BP exchange between the zones in these expressions, and their numerators and denominators are temperature-dependent in the same way, providing the structure of polymer *gs*-micelle–microreactors remains unchanged (i.e. at constant m_s and m). If this is the case, the reaction product composition must also remain unchanged in the temperature range from 329 to 363 K, where the effective activation energy for decomposition (which is actually a sum of E_{0s} for dissociation and ΔH for exchange between the zones) varies from 151 to 126 kJ/mol. The constancy of the composition of the reaction products formed by the aerobic reaction between BP and PS was established by Haas [47 - 49].

REFERENCES

[1] Glasstone S., Laidler K.J., and Eyring H., *The Theory Of Rate Processes. The Kinetics Of Chemical Reactions, Viscosity, Diffusion And Electrochemical Phenomena*, New York, Frick Chemical Laboratory, Princeton University, 1941.

[2] Rado R. and Lazar M., *Vysokomol. Soed.*, 1961, vol. 3(2), pp. 310 - 318. (Rus)

[3] Rado R., Shimunkova D., and Malyak L., *Vysokomol. Soed.*, 1962, vol. 4(2), pp. 304 - 311. (Rus)

[4] Rado R. and Lazar M., *J. Polym. Sci.*, 1962, vol. 62(174), pp. S167 - S168.

[5] Rado R., *Chemicke Listy*, 1967, vol. 61(6), pp. 785 - 806.

[6] Guseva L.N., Mikheev Yu.A., and Toptygin D.Ya., *Vysokomol. Soed.*, 1983, vol. B25(1), pp. 41 - 44. (Rus)

[7] Ledneva O.A., Mikheev Yu.A., Toptygin D.Ya., Gavrilov L.B., and Akutin M.S., *Vysokomol. Soed.*, 1979, vol. A21(6), pp. 1432 - 1437. (Rus)

[8] Nozaki K. and Bartlett P.D., *J. Am. Chem. Soc.*, 1946, vol. 68(9), pp. 1686 - 1692.

[9] Swain C.G., Stockmaier W.H., and Clarke J.T., *J. Amer. Chem Soc.*, 1950, vol. 72(12), pp. 5426 - 5434.

[10] Bagdasaryan Kh.S., *Radical Polymerization Theory*, Moscow, Nauka, 1966. (Rus)

[11] Schwerzel R.E., Lawler P.G., and Evans G.T., *Chem. Phys. Letters*, 1974, vol. 29(1), pp. 106 - 109.

[12] Van Drumpt J.D. and Oosterwijk H.H.J., *J. Polym. Sci., Polym. Chem. Ed.*, 1976, vol. 14(6), pp. 1495 - 1511.

[13] Nakata T., Tokumaru K., and Simamura, *Tetrahedron Letters*, 1967, No. 34, pp. 3303 - 3308.

[14] Saltiel J. and Curtis H.C., *J. Amer. Chem. Soc.*, 1971, vol. 93(8), pp. 2056 - 2058.

[15] Golubev A.N. and Golubeva L.M., *Zh. Org. Khim.*, 1975, vol. 11(7), p. 1566. (Rus)

[16] Madhavan V., Schuler R.G., and Fessenden R.W., *J. Amer. Chem. Soc.*, 1978, vol. 100(3), pp. 888 - 893.

[17] Gragerov I.P., *Uspekhi Khimii*, 1969, vol. 38(8), pp. 1423 - 1447. (Rus)

[18] Cass W.E., *J. Amer. Chem. Soc.*, 1947, vol. 69(3), pp. 500 - 503.

[19] Denney D.B. and Feig G., *J. Amer. Chem. Soc.*, 1959, vol. 81(20), pp. 5322 - 5324.

[20] Hartman P.F., Sellers H.G., and Turnbull D., *J. Amer. Chem. Soc.*, 1947, vol. 69(10), pp. 2416 - 2519.

[21] Hermans P.H. and Van Eyk J., *J. Polym. Sci.*, 1946, vol. 1(5), pp. 407 - 418.

[22] De Tar De Los F., *J. Amer. Chem. Soc.*, 1967, vol. 89(16), pp. 4058 - 4068.

[23] Gill G.B. and Williams G.H., *J. Chem. Soc.*, 1965, pp. 995 - 1002.

[24] Forster W.R. and Williams G.H., *J. Chem. Soc.*, 1962, pp. 2862 - 2873.

[25] Gill G.B. and Williams G.H., *J. Chem. Soc.*, 1965, pp. 7127 - 7136.

[26] Bawn C.E.H. and Mellish S.F., *Trans. Far. Soc.*, 1951, vol. 47(9), pp. 1216 - 1227.

[27] Janzen E.G., Evans C.A., and Nishi Y., *J. Amer. Chem. Soc.*, 1972, vol. 94(23), pp. 8236 - 8238.

[28] Green F.D., Adam W., and Cantrill J.E., *J. Amer. Chem. Soc.*, 1961, vol. 83(16), pp. 3461 - 3468.

[29] Smith W.F. and Rossiter B.W., *Tetrahedron*, 1969, vol. 25(10), pp. 2059 - 2070.

[30] Gelissen H. and Hermans P.H., *Chem Ber.*, 1925, Jahrg. 58, Bd. 1(4), S. 765 - 772.

[31] Cass W.E., *J. Amer. Chem. Soc.*, 1946, vol. 68(10), pp. 1976 - 1982.
[32] Mikheev Yu.A., Guseva L.N., Mikheeva L.E., and Toptygin D.Ya., *Kinetika i Kataliz,* 1986, vol. 27(3), pp. 589 - 596. (Rus)
[33] Mikheev Yu.A., Intern. *J. Polym. Mater.*, 1992, vol. 16 (1-4), pp. 237 - 259.
[34] Mikheev Yu.A., Guseva L.N., Gavrilov L.B., Mikheeva L.E., and Toptygin D.Ya., *Vysokomol. Soed.*, 1984, vol. A26(12), pp. 2618 - 2624. (Rus)
[35] Rogova L.S., Guseva L.N., Mikheev Yu.A., and Toptygin D.Ya., *Vysokomol. Soed.*, 1979, vol.A21(6), pp. 1373 - 1382. (Rus)
[36] Pariyskii G.B., Toptygin D.Ya., Davydov E.Ya., Ledneva O.A., Mikheev Yu.A., and Karasev V.M., *Vysokomol. Soed.*, 1972, vol. B14(7), pp. 511 - 514. (Rus)
[37] Toptygin D.Ya., Pariyskii G.B., Davydov E.Ya., Ledneva O.A., and Mikheev Yu.A., *Vysokomol. Soed.*, 1972, vol. A14(7), pp. 1534 - 1539. (Rus)
[38] Zatsepin A.G., Naymark N.I., and Demina A.I., *Vysokomol. Soe*d., 1976, vol. A18(3), pp. 561 - 566. (Rus)
[39] Wellinghof S.T. and Baer E., *J. Appl. Polym. Sci.*, 1978, vol. 22(7), pp. 2025 - 2045.
[40] Berezin I.V., Denisov E.T., and Emanuel N.M., *Cyclohexane Oxidation,*
[41] Mikheev Yu.A. and Zaikov G.E., *Uspekhi Khim*ii, 2000, vol. 69(3), pp. 249 - 282. (Rus); *Russian Chemical Review*, 2000, vol. 69(3), pp. 231 – 260.
[42] Dorfman L.M., Taub J.A., and Bühler E., *J. Chem. Phys.*, 1962, vol. 36(11), pp. 3051 - 3061.
[43] Nonhebel D. and Walten J., *Chemistry of Free Radicals*, Moscow, Mir, 1977. (Rus)
[44] Guseva L.N., Mikheev Yu.A., and Toptygin D.Ya., *Vysokomol. Soed.*, 1981, vol. B23(5), pp. 360 - 366. (Rus)
[45] Bazhin N.M., *Kinetika i Kataliz*, 1967, vol. 8(3), pp. 532 - 537. (Rus)
[46] Breitenbach J.W. and Frittum H., *J. Polym. Sci.*, 1958, vol. 29(120), pp. 565 - 571.
[47] Haas H.C., *J. Polym. Sci.*, 1959, vol. 39(135), pp. 493 - 499.
[48] Haas H.C., *J. Polym. Sci.*, 1961, vol. 54(160), pp. 287 - 299.
[49] Haas H.C., *J. Polym. Sci.*, 1961, vol. 55(161), pp. 33 - 40.
[50] Russell G.A., J. *Amer. Chem. Soc.*, 1956, vol. 78(5), pp. 1044 - 1046.
[51] Pustoshnyi V.P., Mikheev Yu.A., and Toptygin D.Ya., *Vysokomol. Soed.*, 1983, vol. A25(6), pp. 1295 - 1301. (Rus)
[52] Mikheev Yu.A. and Zaikov G.E., *Chem. Phys. Reports*, 1997, vol. 16(11), pp. 2003 - 2034. (Rus)
[53] Mikheev Yu.A., *Khim. Fiz.*, 1991, vol. 10(5), pp. 715 - 723. (Rus)
[54] Mikheev Yu.A., Guseva L.N., and Toptygin D.Ya., *Khim. Fiz.*, 1987, vol. 6(2), pp. 251 - 257. (Rus)

INDEX

A

B

C

D

E

F

G

H

I

K

L

M

N

O

P

T

U

V

W

X

Y

Z